BEYOND JUPITER

WOMEN'S ADVENTURES IN SCIENCE

BEYOND JUPITER

the story of planetary astronomer

HEIDI HAMMEL

by Fred Bortz

Franklin Watts
A Division of Scholastic Inc.
New York • Toronto • London • Auckland • Sydney
Mexico City • New Delhi • Hong Kong
Danbury, Connecticut

Joseph Henry Press
Washington, D.C.

Author's Acknowledgments

Writing this book was a great adventure, and I thank Heidi Hammel for all she did to make it possible. Heidi repeatedly reminded me that researchers have to let the story emerge from the facts, and that is as true of writing a biography as it is of observing the planets. Many of Heidi's friends and colleagues, too numerous to name, shared brief e-mails. But the most important insights into Heidi's life and work came from these family members, childhood friends, and teachers: Tim Dalton, Irene Dickinson, Jim Elliot, Lisa Gesner, Hazen Hammel, Phyllis Hammel, Michele Jack, Theresa Kennedy, Susan Shaute, and Beth Shaw. Thanks also to the staff of the Institute for Astronomy at the University of Hawaii for taking care of details for my visit to the IRTF, to the researchers there who treated me as part of the team, and to Heidi's children for being themselves when I visited their home. I also want to extend my appreciation to Diane O'Connell, whose skillful editing proved invaluable to the project. Finally, I am grateful to my family, but the one I thought of most while writing this was my late sister, novelist Fay Nedra Zachary, who would have loved hearing about my research, especially those nights on Mauna Kea. Fay, this one's for you! —FB

Cover photo: Planetary astronomer Heidi Hammel is pictured here against a backdrop image of Hale-Bopp, one of the most spectacular comets of the 20th century.

Cover design: Michele de la Menardiere

Library of Congress Cataloging-in-Publication Data

Bortz, Alfred B.
  Beyond Jupiter : the story of planetary astronomer Heidi Hammel / Fred Bortz.
      p. cm.— (Women's adventures in science)
  Includes bibliographical references and index.
  ISBN 0-531-16775-5 (lib. bdg.) 0-309-09552-2 (trade pbk.) 0-531-16950-2 (classroom pbk.)
  1. Hammel, Heidi—Juvenile literature. 2. Planetary scientists—United States—Biography—Juvenile literature. 3. Women astronomers—United States—Biography—Juvenile literature. 4. Women scientists—United States—Biography—Juvenile literature. I. Title. II. Series.

  QB36.H36B67 2005
  520'.92—dc22

                    2005000778

Printed in Mexico.
  2 3 4 5 6 7 8 9 10 R 14 13 12 11 10 09 08 07 06

ABOUT THE SERIES

The stories in the *Women's Adventures in Science* series are about real women and the scientific careers they pursue so passionately. Some of these women knew at a very young age that they wanted to become scientists. Others realized it much later. Some of the scientists described in this series had to overcome major personal or societal obstacles on the way to establishing their careers. Others followed a simpler and more congenial path. Despite their very different backgrounds and life stories, these remarkable women all share one important belief: the work they do is important and it can make the world a better place.

Unlike many other biography series, *Women's Adventures in Science* chronicles the lives of contemporary, working scientists. Each of the women profiled in the series participated in her book's creation by sharing important details about her life, providing personal photographs to help illustrate the story, making family, friends, and colleagues available for interviews, and explaining her scientific specialty in ways that will inform and engage young readers.

This series would not have been possible without the generous assistance of Sara Lee Schupf and the National Academy of Sciences, an individual and an organization united in the belief that the pursuit of science is crucial to our understanding of how the world works and in the recognition that women must play a central role in all areas of science. They hope that *Women's Adventures in Science* will entertain and enlighten readers with stories of intellectually curious girls who became determined and innovative scientists dedicated to the quest for new knowledge. They also hope the stories will inspire young people with talent and energy to consider similar pursuits. The challenges of a scientific career are great but the rewards can be even greater.

Contents

Preparing for the Unexpected

Heidi Hammel is an explorer. As a planetary astronomer she is not actually walking on the surface of other planets, but she is exploring them through images. For Heidi the two giant planets Neptune and Uranus hold a special fascination. Whether she is observing these "ice giants" from one of the world's great observatories in Hawaii or analyzing images that have been collected from a space telescope, she is forever on the lookout for a discovery.

By opening her eyes to small but interesting details, Heidi is always prepared for the unexpected. And when the unexpected comes—even in the far reaches of the solar system—Heidi turns it into an adventure.

Heidi and other scientists once found a giant storm on Neptune that they expected to rage for decades. But only five years later, Heidi discovered it had vanished. Another time Heidi led a team of Hubble Space Telescope scientists who took photos of a battered planet Jupiter as pieces of a fragmented comet crashed into it every day for a week. Heidi's enthusiasm and down-to-earth descriptions of the event made her one of the celebrities of what became known as "the Great Comet Crash."

These discoveries have brought Heidi acclaim, but none of the attention has changed her approach to life in the least. She still looks to the future and sees adventure around every corner, and she wants everyone to come along for the ride.

Heidi turns the wheel
and starts the climb.

Far ahead is
the summit *and one of*
the world's great
observatories.

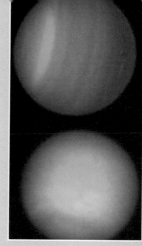

A Date with Two Planets

H eidi Hammel sits behind the wheel of her rental car, preparing to turn off the Saddle Road, which crosses Hawaii's Big Island between two huge volcanoes. She grips the steering wheel a little tighter, anticipating the steep, narrow, curving road that climbs the dormant Mauna Kea [MOW-nah KAY-uh]. The volcano's name means "white mountain," so-called because its peak is often snow covered.

Heidi turns the wheel and starts the climb. Far ahead is the summit and one of the world's great observatories—the Infrared Telescope Facility (IRTF), part of the National Aeronautics and Space Administration (NASA). If all goes according to plan this week in August 2003, she will spend three nights here observing the giant planets Uranus and Neptune. But first she will make a stop at the Onizuka Center for International Astronomy, located 9,300 feet above sea level. The center includes a comfortable dormitory, which will be Heidi's home for the next four days.

The higher Heidi drives, the thinner the air becomes. At an altitude of nearly 14,000 feet, where the Mauna Kea Observatories are located, the air pressure is low enough to activate a passenger airplane's emergency oxygen system. At this pressure, the body isn't getting the oxygen it needs to function properly.

The IRTF *(opposite)* appears to be sitting on top of the world. It's one of several world-class telescopes that share the Mauna Kea summit. Views of Uranus and Neptune *(above)* are shown in false color.

Experienced observers like Heidi always rest for a few hours at the center to give their bodies time to adapt before going higher.

After checking into her room, Heidi calls her husband, Tim. She'd love to say hello to their three children—Beatrix, Tobias, and Lucas—but it's nearly 11 P.M. back in Connecticut. By now they're all sound asleep. Though Heidi has been awake for 18 hours and has traveled 5,000 miles, she wants to hear the details of Tim's Sunday with the children. The conversation is short because Tim has to get up early to go to work the next morning. But as usual, he leaves Heidi smiling.

Heidi's children in 2003 *(from left)*: Tobias (age 3), Beatrix (age 5), and Lucas (age 1).

Then it's time to get down to business. Heidi freshens up and goes to the dining room for a meal with other members of her observing team. Her day is still not over. As the Sun sets, she heads up the nine-mile hill to the IRTF—this time in a four-wheel-drive vehicle. All but the last mile or two is unpaved, with switchback curves. Until recently, it was barely wide enough for two vehicles to pass. Near the top, the road is paved with asphalt to avoid stirring up dust that could damage the observatories' mirrors, lenses, and sensitive equipment. Heidi flicks off her headlights because their glare can interfere with observing. She peers carefully into the darkness as she approaches the facility.

Inside, telescope operator Paul Sears is already hard at work. Tonight Heidi won't be doing any work herself. Her main mission is to give her body time to adjust to the thin air.

She's glad Paul is on duty. A former tour guide, he entertains her with stories of the geology of the mountain, the history of the Big Island, and his many nights operating the telescope. Even with

good conversation, Heidi's long day finally catches up with her. She heads for her vehicle and looks up at the splash of the Milky Way overhead. Mars, glowing deep red, dominates the eastern sky. Uranus and Neptune are in the same general direction but are too faint to be seen by her unaided eyes. Tomorrow night the IRTF Telescope will make them visible.

Heidi drives back down the dark, winding road to the dormitory. It's 10 P.M. when she gets there. She has been awake for 23 hours! As she drifts off to sleep, Heidi thinks about the observations to

Stars of the Milky Way shine above Mauna Kea. Against moonlit clouds stands an "ahu hoku," or star altar, built of rocks and topped by white coral.

come. Neptune is always interesting, but it's a particularly fascinating time on Uranus, the planet with the most extreme seasons in the solar system. Uranus is approaching an equinox—the transition into spring in one hemisphere and autumn in the other. The last time Uranus had similar weather was 42 years earlier in 1961,

The Strange Seasons of Uranus

Uranus takes 84 Earth years to circle the Sun, so its seasons each last 21 Earth years. When Heidi was born, an equinox on Uranus was only five years away. In Earth terms, it was like the beginning of March, the time when early signs of spring appear in the northern hemisphere.

Seasonal changes on Uranus are the most extreme of any planet because its axis—the imaginary line through its north and south poles—has a crazy tilt.

Every planet's axis is angled at least slightly toward the planet's orbital path around the Sun *(below)*. Earth's tilt is fairly large. Its axis is inclined about one-quarter of the way toward its orbital path. But Uranus

spins nearly on its side. Its north pole actually points slightly south of its orbital path!

This makes for unearthly Uranian seasons. The planet's north pole points nearly directly at the Sun during midsummer. Forty-two years later, at the depth of northern winter, its south pole points at the Sun *(right)*.

With such dramatic sunlight shifts, astronomers like Heidi suspect that an equinox on Uranus must be a time of amazing seasonal change. And because it occurs only once every 42 Earth years, it's a once-in-a-lifetime opportunity to discover something interesting.

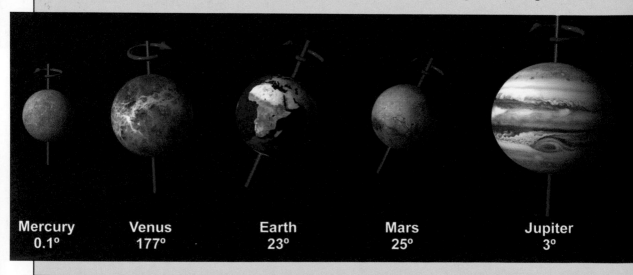

| Mercury 0.1° | Venus 177° | Earth 23° | Mars 25° | Jupiter 3° |

when the young astronomer-to-be was only a year old. No one could have guessed that Heidi would grow up to be an expert on Uranus and Neptune. But then no one has been able to predict most of what has happened in the life of Heidi Hammel—not even Heidi herself.

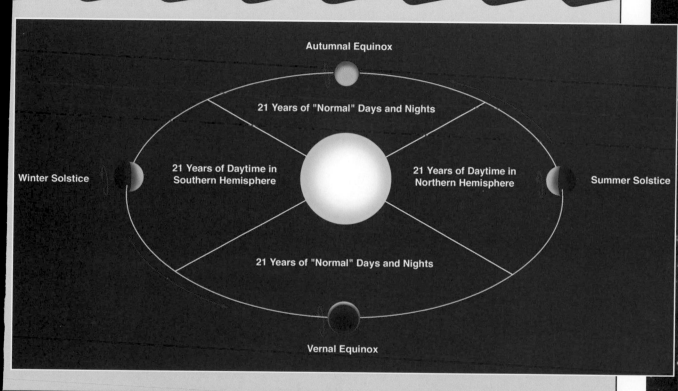

Autumnal Equinox

21 Years of "Normal" Days and Nights

Winter Solstice

21 Years of Daytime in Southern Hemisphere

21 Years of Daytime in Northern Hemisphere

Summer Solstice

21 Years of "Normal" Days and Nights

Vernal Equinox

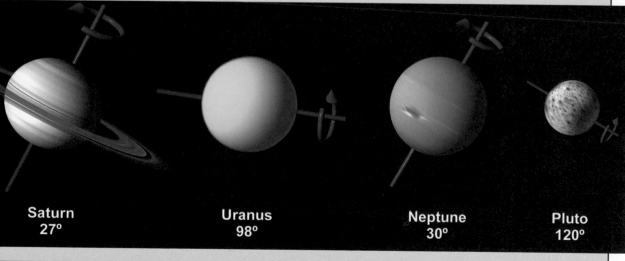

Saturn
27°

Uranus
98°

Neptune
30°

Pluto
120°

Some people might say
that Heidi has been **living**

on **Uranian**
or **Neptunian** time
her whole life.

OPPOSITIONS AND EQUINOXES

2

When people tell Heidi they are jealous of her frequent trips to Hawaii, she laughs. They picture sunny beaches, mountain vistas, and the spectacle of lava flowing into the ocean. But Heidi's visits rarely include any of that. From the moment she arrives in Hawaii, where many of the world's greatest telescopes are located, she's on the job. Her usual routine includes observing the planets from dusk to dawn, then sleeping until early afternoon. She usually works in an office or a library until it's time to get back to the telescope. She grabs meals when she can while fighting off jet lag and the drowsiness and headaches caused by the high altitude.

This down-to-earth toddler playing under the California sun *(opposite)* has a destiny far above the clouds. Saturn *(above)* appears in its golden splendor.

The combination of a crazy schedule and thin air always gives Heidi vivid dreams and bleary eyes. Sometimes the low atmospheric pressure makes her feel bloated and queasy. But she loves every minute of the work, even when everything seems to be going wrong, because she knows that whenever the big telescope is aimed at a new target, she might make a discovery—something nobody has ever seen before.

~ Living on Neptunian Time

Heidi has been on hand for some of the most spectacular planetary discoveries of recent years. Some people might say she has been lucky—and that's partially true. But making discoveries requires more than just good fortune. An astronomer needs to know where, when, and how to look. In Heidi's case, that means allowing the orbits of Uranus and Neptune to determine the schedule of her work, year after year.

A good time to view any planet or asteroid in the solar system that orbits beyond Earth is when that body is in "opposition"—when it and Earth are aligned on the same side of the Sun. The arrangement gets its name because, when viewed from Earth, the other body is in the opposite direction from the Sun. The object is also about as close to Earth as it ever gets. So oppositions give astronomers both their closest views and the most time for making observations. Uranus's and Neptune's oppositions occur about once

Only superior planets—planets that lie outside Earth's orbit, such as Mars (*below*)—can be in opposition to the Sun.

a year, when the faster-moving Earth "laps" them on its path around the Sun. In 2003, both planets were near opposition to Earth in August, so that was when Heidi needed to go to Hawaii.

Some people might say that Heidi has been living on Uranian or Neptunian time her whole life. The day she was born, March 14, 1960, was an excellent time for astronomers to observe those distant worlds. Uranus's opposition was on February 9 that year, and Neptune's followed on April 28. At any hour of the night on March 14, at least one of the two planets was in a good position for observing.

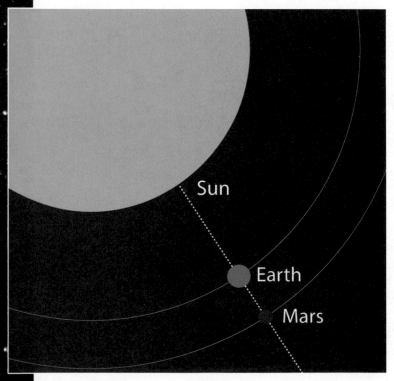

Sun

Earth

Mars

~ Seasons of Change

Even if the world's best telescopes were trained on Uranus and Neptune the night Heidi was born, the images they produced would have shown very little detail. In 1960 the instruments simply were not powerful enough. Neither of those "ice giants" has the colorful bands of clouds that make the pictures of their larger "gas giant" cousins, Jupiter and Saturn, so eye catching.

Without seeing details of the atmospheres, astronomers had no way to observe the weather on Uranus and Neptune. But that didn't stop them from wondering what was happening there. When Heidi was born, the first day of spring, or the equinox, was approaching in the northern hemisphere on Earth. The equinox is one of two days each year when daylight and darkness are equally long everywhere on the planet and a period when the weather is most changeable.

In 1960, the 200-inch Hale Telescope at California's Palomar Observatory was state of the art. It remained the world's largest reflective telescope until the 1990s. Scientists used to have to climb inside it to use it.

The Giant Planets

The planets Jupiter, Saturn, Uranus, and Neptune are called "giants." As the name suggests, they are much larger than Earth. But there is another remarkable difference. These four planets are made up mostly of gases, liquids, and ices. That means they have no solid surface, so even if you could reach these planets by spaceship, you couldn't land on them. Instead, you'd sink through their atmosphere until you were either crushed by the tremendous pressure of the atmosphere above you or melted by the very high temperatures in their deep atmospheres.

Jupiter's 795-mile-thick atmosphere is made up mostly of hydrogen (81 percent) and helium (17 percent), the same gases as the Sun. There are also traces of methane. (Methane is the main component in natural gas.) Jupiter is called a gas giant. It is also the largest and heaviest of all the planets. Its average distance from the Sun is about 484 million miles.

Jupiter

Saturn is the second-largest planet in our solar system. It is almost twice as far away from the Sun as Jupiter. Like Jupiter, its atmosphere is made up mostly of hydrogen (96.3 percent), helium (3.3 percent), and traces of methane. It has a hot, rocky core, and it is also a gas giant.

Uranus and Neptune, too, have atmospheres of mostly hydrogen and helium, but they also have a little more methane. Methane absorbs red light and reflects blue-green light, which is what gives these planets their blue-green colors. They both also have rocky

Uranus

Neptune

cores. Uranus is twice as far from the Sun as Saturn, and Neptune is more than three times as far. These planets are thought to have a layer of ice made up of salty water under their thick atmospheres—that's why they are nicknamed ice giants.

Saturn

The solar system's four bold and beautiful giants are pictured in natural colors by the Hubble Space Telescope, the instrument that changed our view of the universe.

It wasn't just the seasons that were changing when Heidi was born. Her family was also making big changes. Her parents, Phyllis and Robert Hammel, had packed up their belongings the previous June, left their native eastern Pennsylvania, and headed for California. At that time, the Golden State was the Promised Land for young families in search of opportunity, and Bob was eager to start a new job working for the state government.

Phyllis also looked forward to a new life but, much more than Bob, she felt uprooted. Joining a Lutheran church helped her adjust, though she still missed her congregation back in Pennsylvania. Her day-to-day life also changed dramatically. Before the move, she worked as a nurse in a hospital nursery; but when they arrived in California with their four-month-old son, Hazen, and Heidi due in March, Phyllis decided to wait to find work. She wanted to devote her full energies to caring for her young children.

~ Living the American Dream

Two years after Heidi was born, the family was complete with the birth of a third child, Lisa. As older siblings often do, three-year-old Hazen and two-year-old Heidi soon teamed up against the new arrival, throwing toys into her bassinet—until their mother made them understand this was wrong.

This pattern would continue throughout their childhood. Instead of bonding with her sister, Heidi was more likely to do things with her brother. Hazen and Heidi were more physically adventurous than Lisa, so on family outings to places like the beach, they would go exploring with Dad. Lisa enjoyed the quieter company of Mom and playing in the sand.

In many ways the Hammels resembled the idealized suburban American family of that time. Bob Hammel had a career that provided most of the family income, and Phyllis was in charge of the details of raising the children and running the house.

Heidi and Hazen seek out summer fun with their dad.

Heidi, Lisa, and Hazen come together for a holiday photo with a rosy-cheeked and cheerful St. Nick.

Bob was proud to have been the first in his family to earn a college degree, and from an early age his children knew they were expected to do the same. He and Phyllis wanted their children to be well-rounded, so Heidi started swimming lessons at age four and piano lessons at age five. Family vacations were fun, but they were also educational. The Hammels went to museums and traveled to places that taught the children something new about nature, history, or culture.

Bob and Phyllis wanted their children to value reading, so the house was full of books and magazines. They also made sure that learning went beyond books. They encouraged the use of imagination and chose toys and gifts for their educational, as well as entertainment, value. Bob even bought a telescope for backyard viewing, but Heidi, the future astronomer, barely used it. She would rather watch *Star Trek* or study her blueprints of the starship *Enterprise*.

Phyllis especially wanted her children to discover something that was interesting and important in their lives. She considered this the road to success and happiness. "What you have in your head, no one can take away from you," she often told them. "Read. Study. Learn. Dream. Go. Do."

~ Hidden Challenges

From the outside the Hammels appeared to be the ideal American family. But as with all families, there were challenges beneath the surface. Though Bob loved his children, he had a hard time showing it. He was not really a part of Heidi's day-to-day life.

Ironically, the only time she felt close to him was the year he was working on a project thousands of miles away in Thailand, when she was seven. He often sent Heidi and her siblings memorable gifts such as Thai fairy-tale books, kites, and dolls that she kept for many years. He also sent engaging letters that spoke of geckos in his room and exotic foods.

Bob's "hands-off" approach to parenting was not the only problem for Heidi. During her early years, her father changed jobs frequently, and each change meant a move to a new house. Every time Heidi managed to make new friends, the family would be uprooted and she would have to start all over again. The Hammels lived in five homes in three California cities before Heidi was six years old. Then they returned to Pennsylvania and moved twice more before settling in the town of Clarks Summit, Pennsylvania, in 1970.

In the 1960s, souvenirs from faraway places, like this parasol from Thailand, were especially memorable.

Each new position also brought new job pressures, and Bob didn't deal with the stresses very well. Instead, he turned to alcohol to help him forget his problems, at least for a little while. He began to rely more and more on drinking, not realizing he had a condition called alcoholism. People with alcoholism cannot stop drinking—even if they want to—without getting help.

> The Hammels went to museums and traveled to places that taught the children something new about nature, history, or culture.

All Heidi knew was that her father was different from other dads, especially when he had been drinking. She didn't understand why or what was wrong. But she did know she didn't like it one bit.

Long before Heidi first
saw the **swirling clouds**
of Uranus and Neptune,

she was no stranger to
storms.

3

STORMS AND SHELTER

Long before Heidi first saw the swirling clouds of Uranus and Neptune, she was no stranger to storms. It was hard enough facing the emotional and physical changes that every girl experiences in her preteen and early teen years, but Heidi also had to deal with a rigid, alcoholic father.

Fortunately, her mother offered comfort and encouragement, which helped make her home life more stable. Phyllis believed in allowing children the freedom to explore, as long as they knew how far to go. When Heidi heard her mother say, "Don't cross this line," she knew that was the end of the discussion.

Heidi is poised on her monkey swing *(opposite)*, ready for adventure. Stunning sunsets *(above)* are regular events at Mauna Kea.

~ Monkeying Around

As much as Heidi enjoyed exploring her new neighborhood, she longed for the companionship of friends in her new hometown. She didn't have to go far to find one. Theresa Krankel was the same age and lived right across the street. Both girls loved numbers and math. They also enjoyed the same kind of indoor games, especially cards and the board game Monopoly. When they played outside, each displayed high spirits and an adventurous streak—especially when it came to the "monkey swing" in the Hammel's yard.

The monkey swing was a bright red, round plastic seat with a hole in the middle that hung on a long yellow rope from a high, sturdy branch of an oak tree. Heidi and Theresa loved to make that swing live up to its name. They would twist and turn acrobatically, often wrapping the rope around the tree while spinning—sometimes even when swinging backwards.

One day in early summer, Theresa's adventurousness on the monkey swing got her into trouble. She fell off it, landed the wrong way, and broke her arm. For weeks she had to wear a plaster cast and couldn't swim or do other outdoor activities.

Heidi and Theresa spend one of many sunny afternoons together. Monopoly *(below)* was more than a game; it helped Heidi learn how to uncover patterns.

Heidi felt awful. It wasn't her fault, but it was her swing. Besides, she couldn't let her best friend mope around all summer with nothing to do. So the two girls spent most days playing and inventing games. They played with three or four Monopoly sets at once. When they ran out of Monopoly money, they got money from The Game of Life, because that board game had bills up to $100,000.

They made charts and graphs that tracked how many bills they had of each value: $5s, $10s, $20s, and so on. Sometimes they combined card games to make new, more complicated ones, like a way to play gin rummy with five decks of cards.

Though Heidi wasn't aware of it at the time, those summer days were her first scientific experiences. Creating charts and graphs was part of it, but the invented games also taught Heidi to pay attention to details. She learned that discovery often comes from looking at something or asking questions about it in a way that no one else ever had before.

~ Music in Her Soul

While Heidi's intellect was blossoming, her father's alcoholism was taking over his life. Most evenings Bob would come home, plop into his recliner, and drink until he fell asleep, even if the Hammels had visitors—and that was becoming a rare event.

As difficult as this was, Heidi refused to allow her father's problems to dominate her life. Instead, she took her mother's advice to "follow your dreams" and concentrated on activities that enriched her soul. For Heidi those usually included music. She loved all kinds—from classical to pop to rock. She loved listening to music, she loved playing music, and she loved to sing.

When she entered junior high, she became friends with others who felt the same way about music. She met Michele Bendrick and Beth Stauffer. Like Heidi, Michele and Beth were good math and science students, but their love of music drew them together.

> For Heidi, as well as for lots of people young and old, playing music fosters joy and a sense of accomplishment in life.

Beth didn't study piano until seventh grade, but she caught on quickly. A year later, Heidi could tell Beth was "10 times better" than she was, even though Heidi had been playing the instrument since age five. Heidi loved music too much to give it up.

Still, she realized she would never be good enough to make it her career. She looked for instruments that would be more fun and require less practice. She tried guitar and liked it a lot, but she also wanted to play in the high school band or orchestra. Heidi decided that she would try percussion instruments.

At the time, not many girls played in the percussion section, but that didn't bother Heidi. All those years of studying piano paid off—especially with certain instruments. The layout of the xylophone bars is the same as the piano keyboard, so she had no trouble hammering out a melody. Her ear for pitch helped when she played timpani (kettledrums). The other drummers struggled with that because all they knew were rhythm and tempo.

Heidi enjoyed acting, too. In eighth grade she tried out for the junior high play, a comedy called *Rough and Ready*. She won the role of one of a pair of twins whose antics stole the show.

Even when making people laugh, Heidi took her role seriously. She would have been embarrassed to go onstage unprepared. But that is exactly what she was asked to do the next school year at a high school dance called the Freshman Frolic.

Heidi was already feeling self-conscious that evening. Before her date picked her up, she checked herself in the mirror and decided her gown was horrendous. The last thing she wanted to do was show it off to the whole class. When the student emcee called on her to entertain everyone with a scene from *Rough and Ready*, she shook her head. She said she hadn't played the part in months, but her classmates egged her on.

The emcee practically dragged Heidi onstage. She was mortified, but she managed to keep her cool. She flubbed some lines, but the audience laughed and applauded. Not even Beth or Michele could sense her embarrassment. She didn't even want to show them how upset she was.

Abington Heights Junior High School

PRESENTS

Rough 'n' Ready

FRIDAY EVENING   March 21, 1975

CURTAIN 8:00 P.M.

Adult Ticket $1.50

№   78

~ *Blackout!*

Heidi was becoming skillful at hiding her emotions. Too skillful, in fact. At home she barely discussed her feelings with her mother, though the tensions in the Hammel household were gnawing at everyone. But keeping everything bottled up inside was not healthy. It affected Heidi's appetite and she wasn't eating well. Phyllis didn't make an issue of it until one Sunday morning when Heidi fainted while singing in the church choir.

The doctor blamed her fainting on severe stress. For Phyllis the message was clear. She had been denying the effects of Bob's alcoholism on the family for too long. She now knew she had to make a major change in their lives for the sake of the children. Since Bob couldn't stop drinking, she would have to end the marriage.

It didn't take Phyllis long to realize she had made the right decision. When she told the kids that she and their father were getting divorced, Heidi was not surprised. The first question she asked was, "When does he move out?"

Hazen and Lisa felt the same way. The three Hammel children sometimes missed their father, but mostly they were relieved not to be dealing with his unpredictable behavior. And good things started to happen. People came to visit. All of a sudden, neighbors and relatives felt welcome, and Heidi and her siblings weren't ashamed to invite their friends over. They finally felt like they could relax in their own home.

Band mates Heidi and Michele pose for a quick "off-duty" snapshot.

~ Finding One's Space

With her home life at last settling down, Heidi became immersed in school life. At her high school, Abington Heights, some of the students divided into band kids, cool kids, and jocks. Hazen and Heidi were band kids—smart, fans of science fiction, and "pretty nerdy" in the eyes of Lisa and the rest of the students. They didn't care much about weekend parties or wearing the latest clothes. Lisa, on the other hand, was athletic, involved in drama, and part of the cool popular crowd.

But the differences between the teenage sisters went beyond what they did for fun and the people they hung out with. For years they had to share a bedroom, but by the time Heidi started high school, that had become a recipe for disaster. Lisa liked her room to be neat as a pin, while Heidi threw everything everywhere. They divided the room in half and hissed, "Don't let your stuff go across this line!" But nothing worked, and they were always squabbling.

Phyllis got tired of listening to their arguments and came up with an unusual solution. She separated the sisters and put Heidi in a room with her brother. Heidi had her mess, and Hazen had his mess. So what? They each knew where their stuff was and didn't care what anybody thought. So for most of high school, Heidi and Hazen shared a room full of organized chaos. It worked because they respected each other's privacy. Heidi and Hazen also shared an interest in music. They both sang in the concert choir and the madrigal choir. In addition to the band, Heidi played in the orchestra. One year she was selected for the regional band and was astonished to be chosen as one of two percussionists for a featured marimba duet.

Heidi (front row, middle), an alto in high school, stands proudly with her fellow concert choir members.

~ Doing It Right

Susan Shaute didn't just teach her students how to perform in the arts. She taught them how to perform in life.

Of all Heidi's teachers at Abington Heights, the one who had the most influence on her was choral director Susan Shaute. Miss Shaute was a former Rockette at Radio City Music Hall in New York City, and she believed in and demanded professionalism from her students. Every choral lesson started with the same exercises that professional singers did. "This is not Amateur Night in Dixie," she often told

them. "If you're going to do a job, do it right. Don't do it halfway," she insisted. "Always strive to be professional."

Heidi took those words to heart. She was eager to become a professional—but a professional what? She had no idea! She supposed that going to college would help her figure that out. Everyone expected her to go into music, but she didn't feel it was right for her. Everyone knew she was smart, but what colleges would suit her best? Which colleges could her family afford?

It was her senior year of high school and time to make an important choice. Choosing and getting into the right college was Heidi's next job, and she was not about to do it halfway.

Senior class pictures document an especially important life transition for high school students; it was no different for Heidi.

Maybe she ought
to consider **science**
—she did like it in a way—and if so,

why not apply
to a top school like MIT?

THE ROAD TO MIT

Heidi's father was always proud to say that he graduated from the Pennsylvania State University, better known as Penn State. Besides being convenient to Clarks Summit, it offered an excellent education at an affordable cost for state residents. Most important, Penn State students could major in many different fields. It was no surprise then that Heidi set her sights on her father's alma mater.

In the spring of 1978, MIT was only a summer away for Heidi, shown here as a graduate of Abington Heights High.

~ A Change in Plans

Heidi was accepted easily at Penn State. She was barely into her senior year of high school when her plans were set. Her father wasted no time sending in a deposit to hold Heidi's spot.

Heidi was satisfied with her choice until one day in calculus class. Because it was an advanced course with only four students, the teacher, Mary Rhodes, got to know each of them well. She wanted to make sure they were all heading to the right schools. "Where are you all going to college?" she asked.

"I'm going to Penn State," Heidi announced when it was her turn.

Like all good mentors, mathematics teacher Mary Rhodes opened Heidi's eyes to options she had never considered.

Mrs. Rhodes responded, "Have you thought about anything else?"

Heidi was puzzled. Why go anyplace else? "We live in Pennsylvania. My dad went to Penn State . . ."

Mrs. Rhodes knew that Heidi was a strong math student and that she also did well in her science courses. She interrupted Heidi and asked, "Well, why don't you apply to MIT?"

MIT—the Massachusetts Institute of Technology— was and still is one of the very best schools in the country for science and engineering. But up until that moment Heidi had never heard of it. Suddenly Penn State did not seem like such an obvious choice. Maybe she ought to consider science—she did like it in a way—and if so, why not apply to a top school like MIT?

Heidi filled out an application and thought about which teachers might write letters of recommendation. Who should she ask? Of course, Mrs. Rhodes happily agreed to write one. Then, because MIT was a science college, Heidi went to her chemistry teacher. She had done well in his class and was known for being a bright, serious student. She didn't expect he'd have any problems with writing her a strong recommendation.

"Can you write me this letter of recommendation?" she asked.

"No," he replied.

"Why not?" Heidi persisted.

"You'll never get into MIT," he said. Clearly, he wasn't hiding his old-fashioned belief that girls weren't "smart" enough to get into science colleges.

"Oh." Heidi responded. How could he be so sure? But she knew the chemistry teacher had his mind made up. She asked her history teacher for a recommendation instead.

A few months later an official acceptance letter from MIT arrived in Heidi's mail. She couldn't wait to show it to her teachers. She wished she could shove it into the chemistry teacher's face, but she simply said, "Look, I got into MIT."

Heidi has never forgotten his response: "It's only because you're a woman. They have quotas to fill."

She was appalled. *Fine,* she said to herself, *if that's how you feel. Great. See ya later!*

~ Tears—Then a Road Trip

Getting accepted to a school like MIT is difficult, but getting a scholarship is harder. Heidi was not offered enough financial aid, and there was no way her parents could afford the rest of the school's high tuition, room, and board. So she went to the guidance counselor at her school to ask for help.

The counselor called MIT and put Heidi on the phone. Heidi described her family's finances and told the MIT representative how much she wanted to go there. Still, it had no effect. The representative simply replied, "I'm sorry, but you can't afford to come here." Heidi broke into tears.

> Heidi filled out an application and thought about which teachers might write letters of recommendation. Who should she ask? Of course, Mrs. Rhodes happily agreed to write one.

Heidi may have cried, but she was not about to give up. When she told her family about the conversation, her brother, Hazen, said, "Just go up to MIT and talk to them." He was only half serious, but he knew that MIT was a wealthy school with plenty of financial aid money.

Heidi thought that was a fine idea. She drove five hours to Cambridge, Massachusetts, where MIT was having an open house called Technology Days. She poked around the campus a bit and liked what she saw. So she marched into the financial aid office and asked again in person what she had previously asked on the phone.

On warm, sunny days, the expansive lawn surrounding MIT's Great Dome provides a relaxing meeting place for busy students.

Faced with such a determined student, the financial aid officer called Heidi's father and persuaded the loyal Penn State graduate that his daughter would be better off attending MIT. The school offered some financial aid, but it would still cost him much more. Heidi would also have to do her part by working part-time on campus. In the end her father supported her choice and agreed to come up with the extra money. Heidi looked at her options: comfortable Penn State, where she knew she would do well, or MIT, where her future looked less certain but more interesting. Her choral director, Miss Shaute, had taught her that professionals thrive on challenges, so for Heidi the choice was clear.

After she settled on MIT, Heidi faced the usual next questions from people: "What are you going to major in?" and "What do you want to do when you graduate?" She didn't have an answer to either one.

Heidi still hadn't set specific goals when she arrived at MIT as a freshman in the fall of 1978. Phyllis went along to help her daughter settle in, and she stayed long enough to go through a reception line to meet the school's president. When he asked Heidi what she planned to study, she replied, "Oh, I don't know what I am going to study. I just know this is the best place to learn, and I came here just to learn."

If she could have read her mother's mind at that moment, she would have heard, *Oh, Heidi! You have to have goals! You can't say things like that!* But the president had a different reaction. He smiled at Heidi and responded, "Well, with that attitude, you're going to do very well here among us."

A few months later Heidi wasn't sure she agreed with that prediction.

The nearly straight-A student
who had **breezed**
through high school

was being **challenged**
in a way that
she had never been challenged before.

CHARTING A PATH

5

For a freshman entering MIT, having a major was not important. Except for one or two electives, everyone took the same courses. And that "everyone" included some of the smartest people Heidi has ever known.

High school had been easy for Heidi. At Abington Heights she coasted through almost every class, got straight A's or an occasional B, and had plenty of time left for "a life," including music, school plays, stargazing, or just hanging out with friends.

But at MIT everyone was as smart as Heidi (or smarter), and the courses were harder than she imagined. No coasting here! "MIT was just an awful, awful experience for me," Heidi now recalls. "You get to MIT and you work and work and work and then you fail. And work-work-work harder, and you fail." As a freshman, she did well in only one course: calculus. "But physics, failed-failed-failed, couldn't do it at all." The nearly straight-A student who had breezed through high school was being challenged in a way that she had never been challenged before.

Still, Heidi wasn't about to give up. Somehow, after she "failed-failed-failed" some tests, she just did what she had told the president she would do in that new-student reception line. She learned until she could pass the required physics courses. After that it was just as her mother had said. Once she had something in her head,

Here we go! Heidi arrives in Cambridge, Massachusetts, and begins life on campus. Reading *Scientific American* magazine *(above)* is just one way to keep up on science and technology topics.

no one could take it from her. And once she had something in her head, the grade she got in the course didn't matter to Heidi anymore. By the end of her freshman year, Heidi did pass all of her courses, though, as she says, "by the skin of my teeth."

~ Astronomy Beckons

In her sophomore year, Heidi had room for a few electives. She looked at the course offerings, and two caught her attention. One was a history course and the other was a new observational astronomy course. The astronomy course was a technical, hands-on, learn-by-doing course, the kind she liked best. Besides, though she knew little about astronomy, she thought it might be fun.

At the first meeting of the astronomy course, Heidi discovered she was the youngest of four students in the class. Two of the others were graduate students, and one was a senior. That was not a good sign for a sophomore who had barely gotten a C in freshman

Dr. James Elliot examines a printout of the rings of Uranus in 1977.

physics. But the professor, James Elliot, assured Heidi that she was exactly the kind of student he hoped would sign up.

MIT had an off-campus student observatory that housed a small telescope. For one class project, students had to use the telescope to solve a research problem. Heidi decided to take photographs of the asteroid Pallas. Asteroids are rocklike masses mostly found between Mars and Jupiter. Astronomers refer to them as "minor planets."

Like planets, asteroids move across the pattern of the stars. Heidi wanted to see how well she could use photographs of Pallas taken at different times to determine its orbital path around the Sun. Since the orbit of her asteroid was well known, she could

compare her calculations with the
published orbit. It wasn't original research,
but it was certainly good training. It was a chance
for Heidi to learn how to use a telescope and how
to read the sky.

All Heidi needed was for the weather to cooperate. But finding
clear evenings to observe was a tall order in New England,
especially for a student with little time to spare. Heidi was juggling
a heavy course load and a job to pay some of her tuition. It didn't
take her long to realize she had bitten off more than she could
chew. She needed to drop a course, and it had to be one of her
electives.

It was a difficult choice. She decided to talk to Dr. Elliot about it.
She hesitated as she knocked on his office door late one afternoon.

The professor invited Heidi in. "I have to drop either astronomy
or my history course, because I just can't keep both going at once,"
she explained. "And the skies have been so awful that I'm not
getting any data! I think I'm going to have to drop your course."

Heidi's hands were clenched so tightly that her knuckles were
white. Dr. Elliot looked at her and then went over to the window.
"It looks clear," he noted. "Let's go out and see if you can get the
data you need tonight. If that works, you won't have to drop
the class."

Pallas, second largest
asteroid in the solar
system *(top)*, is imaged
over several hours by
the Doane Observatory
at Chicago's Adler
Planetarium. This
color-enhanced image
of asteroid Ida *(above)*
was recorded by the
spacecraft *Galileo*.

In the car, Heidi and Dr. Elliot reviewed the procedures they would follow at the observatory: Aim the telescope in the direction of Pallas, attach a Polaroid camera to the eyepiece, and take a picture. Then wait about 15 minutes and take another picture. Do this over and over, labeling each picture with the date and time. Don't spend too much time studying the pictures. Just gather the information—data—and analyze it later.

Heidi photographed Pallas from here, the off-campus Wallace Astrophysical Observatory. It is named after George R. Wallace, Jr., a member of MIT's class of 1913 who supported its construction.

~ "Too Much Fun"

As soon as she and Dr. Elliot got back from the observatory, Heidi announced to him, "I can't drop this class. It's too much fun." Fortunately, the weather that night had been clear, and they got plenty of good data to work with. Each picture showed the same pattern of stars, except for one object that moved slightly between each exposure. That was Pallas, and Heidi knew she would be able to measure its changing position, tabulate the results, and calculate its orbit.

In some ways the process was not very different from the monkey-swing summer of making charts and analyzing graphs with Theresa Krankel, but the results were much more exciting.

Good fortune was certainly at work that night, but so was Heidi's good judgment. Faced with a choice, she allowed astronomy one more chance and discovered much more than she ever expected.

Not only did she determine the orbit of an asteroid, she also discovered her own path through the MIT universe. Before the semester ended, she decided to major in astronomy and allow her mother's words to guide her. She would read about stars and planets and galaxies and more. She would study what others already knew about them so she could learn more on her own. She would allow herself to dream, because imagination opens a person's eyes to new ways of seeing. And with vision guided by imagination, she would go out into the world and do what none had done before her.

From then on, Heidi knew she belonged at MIT. The courses were still difficult, and she still had to work-work-work at what she called "a hideous pace." But when people back in Clarks Summit asked her what her major would be, she had an answer.

~ A Sudden Move

Even while suffering through the worst times with her classes at MIT, Heidi managed to find time for a social life. She began a serious relationship with a young man whom she met as a freshman, Olin Harbury. With Olin, Heidi discovered—and became passionate about—a musical group called the Grateful Dead.

The Dead was an unusual band with a

The Grateful Dead performs for thousands of fans at the Greek Theater in Berkeley, California, in 1984.

remarkable following. Although people could buy recordings of their songs, a true fan had to experience a live concert.

True Deadheads would set aside everything else at least once in their lives to follow the band on tour, as Heidi and Olin did on spring and summer breaks when they didn't have classes. In addition to sharing her love of the Dead, Olin was, in Heidi's words, "a very wonderful person." That, it turns out, was the problem.

When Heidi went home for winter break her junior year, she was glad to be away from the struggles of her courses. As usual, her father was invited for Christmas Eve, and, as usual, he drank excessively. Heidi could see he was still in denial about his alcoholism. Instead of getting treatment for his alcohol addiction,

Music once again became Heidi's lifeline, and the guitar was her connection to a new circle of friends.

he had begun seeing too many doctors who prescribed too many medications to treat his depression. The combination of pills and alcohol was a dangerous mix, which led to his passing out in one of Phyllis's upholstered chairs after dinner. He was still in that chair when Heidi awoke the next morning. Finding her father passed out yet again—and this time on Christmas morning—was the breaking point for Heidi. She'd had enough. She was so appalled that she bolted from the room, packed up her car, and drove back to MIT.

Everything was closed at the school, and the only students around were those from overseas who couldn't afford the long trip home. That suited Heidi just fine. She wouldn't have to tell anyone what happened, especially Olin, whose dorm room was right next to hers. She wasn't sure why, but she felt something was wrong with her life—and maybe with herself. She couldn't inflict whatever that was on a wonderful guy. A room opened up in a different section of the building, and she moved out before Olin got back to campus. She ended their relationship without telling him why.

Music once again became Heidi's lifeline, and the guitar was her connection to a new circle of friends. They would get together in the dormitory and play bluegrass, a style of country music played on stringed instruments. Still, it wasn't hard to see that something was bothering the usually outgoing Heidi. One of the group, Jeff Menoher, decided that she needed to break out of her shell.

Without asking her, he signed them up to play at the Mezz, a coffee shop at the MIT student center.

"What are you talking about?" she shrieked when he told her. She couldn't help thinking of the Abington Heights Freshman Frolic. "I don't do that kind of thing. I don't play guitar in public."

"You don't have a choice," Jeff insisted, thrusting a piece of paper at her. "Look, here we are on the schedule. We have to do it."

So Heidi reluctantly agreed. The group was a hit, and that night at the Mezz turned out to be the beginning of a part-time bluegrass career for Heidi.

More important, it helped her reconnect with life at a difficult time. Heidi was a musician and an astronomer-to-be. She was also the daughter of a man heading for self-destruction. All of this would come together a few years later, thousands of miles away from her life at MIT—and in one of the most beautiful places in the world.

Heidi played percussion for the MIT Musical Theatre Guild's production of *Fiddler on the Roof.* The Guild, completely student run, is the oldest and largest theatre organization at MIT.

She soon discovered that her *skills* went

Research Assistant

MIT undergraduate student Heidi Hammel has been appointed as a summer research assistant at Lowell Observatory. Hammel (left), is assisting Nathaniel White, Ph.D., astronomy, a senior staff scientist, with lunar occultation observations and data reductions.

(Observatory Photo)

MIT StudentUndergraduate Assisting at Observatory

What do the Massachusetts Institute of Technology and Lowell Observatory have in common? Her name is Heidi Hammel, an undergraduate student majoring in planetary science and astronomy at MIT, Cambridge, Mass.

the Lowell Observatory. "The summer appointment will allow me to gain additional hands-on astronomical research in a new and different place while learning about lunar occultations. As an added advantage, I am making contacts

beyond numbers and

mathematics *courses*.

6

BECOMING AN
ASTRONOMER

To earn a bachelor's degree in earth and planetary science from MIT, Heidi had to take physics courses—lots of them—and she continued to struggle. She got mostly low grades, but she managed to pass. As much as she hated the intensity of her courses, Heidi loved the work of astronomy. During her last three years studying at MIT, Heidi became Dr. Elliot's student assistant and helped him on a number of different projects.

She soon discovered that her skills went beyond numbers and mathematics courses. Science is built on measurements—data—and Heidi had a knack for understanding what the data was telling her about the phenomena she was observing. And if the observations weren't as clear or as detailed as she would have liked, Heidi even came up with ways to improve the instruments.

For instance, she found that the tube for a 16-inch telescope was too short. Light was coming into the telescope at angles it wasn't supposed to. So she designed an extension for the tube that was just long enough to cut off stray light but not so long that it unbalanced the telescope. To fashion this "light baffle," she had to take measurements, do calculations, learn how to craft something out of metal and bolts and screws and then attach it to the telescope. The apparatus is still in use at MIT today.

A summer internship at the Lowell Observatory in Arizona offers Heidi *(shown opposite with Dr. Nathaniel White)* valuable real-world experience. Bachman Hall at the University of Hawaii *(above)* houses offices for the president and staff.

She also oversaw the redesign of a brightness-measuring instrument called the "black photometer" on the 24-inch telescope. This instrument measures light from astronomical sources, such as stars or planets. But the photometer was built to measure light at only one wavelength. Heidi helped redesign it so that a filter could be inserted that would allow different colors of light to come through. This change made the instrument more useful because it could detect more wavelengths.

Heidi found Dr. Elliot's research fascinating, especially his best-known work with a technique called stellar occultations. Occultations occur when relatively nearby bodies, such as planets or asteroids, pass between Earth and a star, blocking or dimming the star's light. Just as astronomers can predict the times and places that solar eclipses are visible as they sweep across the Earth, they can also predict the times and places that occultations will occur. If they can get to the right place at the right time, they can measure a shadow cast by starlight. From the way the star's

Occultation in progress: As a planet passes between Earth and a star *(left)*, the planet's atmosphere causes the star to appear dimmer *(center)*. At occultation *(right)*, the view of the star is blocked by the planet.

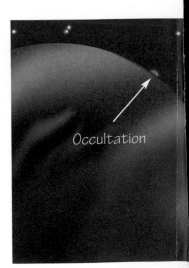

Occultation

light changes, astronomers often discover something about the body that is moving in front of it.

This technique led to Dr. Elliot's most famous discovery. In 1977, he and his colleagues flew in a special airplane equipped with a telescope to follow the faint star shadow of Uranus as it swept across the sky. Sooner than they expected, the starlight

Waves of Light

Light travels through space as a wave of electromagnetic energy. Like a wave in the ocean, an electromagnetic wave has peaks and valleys that can be closer together or farther apart. The spacing between its peaks is called the wavelength. Scientists use the length of the waves to distinguish between different kinds of electromagnetic radiation. Radio waves have the longest wavelengths, followed by microwaves, and then infrared radiation.

Visible light—the light we see—comes next. The different wavelengths of visible light make up the colors of the rainbow, sometimes called the visible spectrum. Deep red has the longest wavelength of all the colors we can see. As the wavelengths get shorter, our eyes see orange, then yellow, then green, then blue, then violet (and all the colors in between).

As the wavelengths get even shorter, they form ultraviolet light, X-rays, and gamma rays, none of which we can see. But astronomers can use these wavelengths to explore the universe.

A telescope works by gathering and focusing dim light—visible or invisible—so it can be "seen." The telescope's mirror gathers light from a planet or other distant object. The mirror concentrates the light at a focal point, where astronomers often put various devices, such as filters that allow only certain colors to pass through. By learning what wavelengths of light are absorbed or reflected by an object, such as a planet, scientists can determine what elements and compounds make up the atmosphere of the planet.

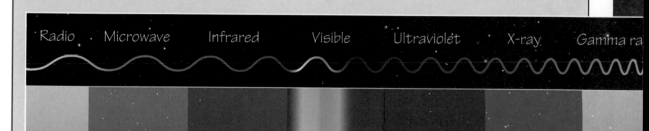

Centuries ago, astronomers like Galileo Galilei and Nicolaus Copernicus were limited to using only the visible light portion of the electromagnetic spectrum (shown above). Today astronomers have many more options. Heidi images mainly in the infrared. Some astronomers study objects in space that emit (give off) strong radio waves, such as sunspots or solar flares. Ultraviolet light is used to detect the chemical composition, densities, and temperatures of interstellar gases. High-energy waves like X-rays and gamma rays can reveal some of the more energetic aspects of the universe, like exploding stars, galactic collisions, and black holes.

The Kuiper Airborne Observatory *(left)* flies with its infrared telescope hatch open. High above Earth's atmosphere, scientists *(right)* can gather important astronomical data unobtainable from the ground.

False-color images can help highlight planetary features, such as the Uranian ring system, an unexpected and stunning discovery in 1977.

began to dim, then brighten, then dim, then brighten again several times before the planet blocked the star completely. As the full occultation ended, the star began winking again. They were seeing the silhouette of Uranus's rings—something that had never been observed before. It was the very first evidence that any planet other than Saturn had rings. (Today, scientists know that all four giant planets—Jupiter, Saturn, Uranus, and Neptune—have rings.)

As Dr. Elliot's assistant, most of Heidi's work involved occultations by the giant planets, which often produced information about those planets' temperatures, pressures, and atmospheric compositions. That work, plus several summer internships, launched Heidi on the path to becoming an expert observer by the time she graduated from MIT in 1982. She was now ready for the next step toward becoming a professional astronomer. She would head to graduate school.

~ Aloha, Neptune!

To get into graduate school, Heidi had to take some standardized tests called the Graduate Record Exams (GREs). Not surprisingly, her GRE score in physics was low. Still, being a graduate of MIT and having a strong recommendation from Dr. Elliot, Heidi's application for graduate work was accepted by the Institute for Astronomy at the University of Hawaii.

The astronomy program at the institute was very strong. Its facilities included two of the world's most respected observatories— Mauna Kea Observatory and Mees Solar Observatory. But Heidi had another reason for wanting to go there. After four difficult years at MIT, she wanted to move as far away as she could and never come back. And she couldn't get much farther away than Hawaii without leaving the United States.

But once she got there, the novelty of being more than 5,000 miles from MIT soon wore off. Heidi realized that she was in a very small state far from the people she needed most. Astronomy was important to her—even exciting—but her life was not complete without family or music and people to share it. Sometimes she felt as isolated as if she had really traveled to Neptune instead of to a place to observe it.

Fortunately, one of her office mates, a young man named Dana Backman, sensed Heidi's mood. When he found out she was Lutheran and could sing, he knew just what to do. "I'm going to call the church choir where I go, because they're looking for singers," he told her. Listening to his end of the conversation, Heidi couldn't help but think of how Jeff Menoher had signed her up to play at the MIT Mezz.

University of Hawaii students stroll along tree-lined McCarthy Mall, located on the school's Manoa campus in Oahu.

"Hi, Carl? My new office mate here—she sings! Yeah, I think she'd sing in the church choir. I'll send her down on Thursday night."

Suddenly Heidi was singing with the Lutheran Church of Honolulu choir. She connected immediately with its director Carl Crosier and his wife Katherine. Like Miss Shaute at Abington Heights, they didn't accept less than the best from anyone.

It wasn't long before Heidi was totally immersed in the church and its music. Her guitar-playing made her the ideal person to lead the congregation's contemporary service. Every Sunday she arrived at 7 A.M. to set up the early service. She stayed to sing in the choir for the more formal second service. By the time she left, it was usually close to 1 P.M.

Heidi no longer felt isolated. The church offered her a sense of fellowship, and the prayers and melodies reminded her of worshipping with her mother back in Pennsylvania.

The church connection was never more important in Heidi's life than on July 18, 1983, about a year after she arrived in Hawaii. Her father had managed to stop drinking for a few years. Then some

On her last day with the Lutheran Church of Honolulu choir, Heidi poses with her colleagues in song, Carl Crosier and Sissel Sodal, a soprano from Norway.

old friends invited him out to celebrate the Fourth of July. An alcoholic is never more than one drink away from a relapse, and he began drinking again. Two weeks later he died of an overdose of prescription medications and alcohol.

Whatever Heidi's feelings about Bob Hammel, he was her father. She knew he loved her and wanted her to succeed, and he had seemed to be trying to get his drinking under control at last. Now that she was beginning her professional career, Heidi had been looking forward to telling him about it, one adult to another. Her opportunity to do that died with him. Heidi had never felt so far from home. She needed to be with her family, but she couldn't afford plane fare for the 10,000-mile round-trip. Instead, she turned to her church family.

The church offered Heidi what she needed most: community, comfort, and the strength to face her loss. It also gave her opportunities to do meaningful work and provided a stable foundation for the ups and downs of her life.

The Dreaded "Quals"

Heidi's connection with the church and music helped take some of the pressure off her adjustment to life as a graduate student. She quickly discovered that graduate school is very different from college. She intended to earn a doctoral degree, the highest academic degree that is awarded. To begin, Heidi needed two years of advanced courses in all aspects of astronomy. Then she had to come up with an idea for a research project to explore an important scientific question that no one had ever answered. For the grand finale, Heidi would have to complete the project and write a thesis, a book-length scholarly report on her results.

As if coursework and thesis research weren't enough, Heidi also had to clear a hurdle at the end of her second year. She had to pass a written test on every area of astronomy and then pass an oral exam, during which a committee of professors could ask her anything they wanted to. This process was called a qualifying examination.

> She quickly discovered that graduate school is very different from college. She intended to earn a doctoral degree, the highest academic degree that is awarded.

For Heidi and the five other students who took the "quals" at the same time she did, the exams were a daunting ordeal. But while the others all passed, the faculty committee told Heidi she would have to try again. She would be given a second and final chance a few months later. In the meantime, she could start on her research. But she wouldn't be considered an official doctoral student until she passed the quals.

Heidi was so upset that she went home and cried the whole night. Still, thanks to MIT, she knew how to deal with failure.

The next day she analyzed the situation to figure out what had happened.

Unlike her struggles with physics, this time she was confident she knew the course material well enough to pass. But she knew there was something different about the oral exam. Faculty members sometimes failed a graduate student on the orals to send a message. But what message might they be trying to send her?

She finally decided it had more to do with her attitude than her knowledge or ability. From the day she first set foot in Hawaii, she had approached graduate school in an unusual way. Usually new graduate students are surprised by the difficulty of graduate school. But after her MIT experience, Heidi was telling her classmates, "Give it a rest. This is a holiday. What do you mean, hard work?"

Most of the other astronomy students did everything together: worked, rented houses, and socialized. Astronomy was their whole life. Not so for Heidi. She was never happy doing only one thing. She needed to be part of the broader community, and she needed her music.

So Heidi worked hard on her courses, but mostly on her own. She preferred to live with students from other academic departments. And she spent time looking around for other people who wanted to play bluegrass

Graduate school is tough work, so Heidi stayed plugged in to music, whether getting down with bluegrass or listening to her favorite band.

music with her. She also maintained her Deadhead status by flying to San Francisco once a year for Grateful Dead concerts.

None of that should have mattered when it came to judging her qualifying exam, but she suspected it did. Although she was probably working as much as her classmates, she thought it might have looked as if she wasn't as committed to the program as they were. The attitude around her was, *"You've got to be focused, narrow, and that's your life."*

That single-mindedness fits most people's images of scientists, but it didn't suit Heidi at all. She needed more than astronomy and was not about to change her personality, but she decided to change her approach to work. She started showing up at 7:30 A.M., before most of the other graduate students and faculty arrived, and she stayed until 5:30 or 6 P.M., when nearly everyone had gone home.

From the day she first set foot in Hawaii, she had approached graduate school in an unusual way.

After 6 P.M., that was it. Her life outside astronomy didn't change after failing the quals. She still hung out with musicians and even played "gigs" on the streets of Waikiki, but she simply stopped telling her classmates about any of it.

She didn't plan to say much to her professors either. She could see a positive shift in their attitudes already, and everything was on schedule for her upcoming research observations of the atmospheres of Uranus and Neptune.

~ *The Mythical Phoenix*

One day Heidi got a call from one of her musician friends telling her about a professor named Elizabeth Wichmann. Dr. Wichmann had translated some classic Chinese operas into English, and she needed string players for a performance of *The Phoenix Returns to Its Nest*.

Heidi protested, "I don't know anything about Chinese music."

"That's okay. They've got teachers here from China," her friend urged. "They'll teach you how to do it."

Well, it might be an adventure, Heidi decided. "Sure, I'll try anything once!"

When she arrived at the tryouts, Heidi was pleased to discover that Dr. Wichmann needed percussion players, too. Now that really seemed interesting! All that work with the Abington Heights High School band and orchestra was about to pay off. Suddenly, for months on end Heidi's nights were filled with opera rehearsals.

Naobo (cymbals) and xiao luo (gong) are percussion instruments used to punctuate action and emotion.

Every day she plowed away at astronomy from early morning until 6 P.M. Then she would leave, eat dinner, and go to Chinese music classes. She didn't tell anyone in the astronomy department about it because she was concerned they would think she wasn't taking her main job seriously enough.

When the time of the public performance came, she couldn't keep her participation a secret any longer. She told some people on her thesis committee, who eagerly attended the performance. They were astounded and impressed when they saw her onstage playing clangy cymbals called naobo

The actors, musicians, interpreters and teachers of the Jingju (Chinese opera) production of *The Phoenix Returns to Its Nest.*

[NOW-bo] and a little gong called a xiao luo [she-OW lo] that makes what Heidi calls "this really bizarre sound. It goes *doi-oi-ing, doi-oi-ing, doi-oi-ing* if you hit it right. If you hit it wrong, it goes *pa-ack*."

Wearing celebratory wedding gowns, Lynn Weber and Kyle Kakuno play the roles of a young heroine and hero.

But that was only the beginning. Dr. Wichmann sent videotapes of the performance back to China and got an invitation to tour and perform the opera there in English. Heidi wanted to go, but she knew it would take her away for a full month during prime observing time for Uranus and Neptune.

When she told her thesis advisor and a key committee member about the invitation, they didn't hesitate. "This is an opportunity you can't miss," they told her. "Go and do this."

When Heidi came back from China, she felt like the mythical Phoenix bird that died in flames and arose from its ashes. Uranus and Neptune were still in good position for observing, and she was eager to retake the qualifying exam. This time she was sure that she would pass.

For Heidi, **planets**
offered the chance

to feel like an
explorer.

7

NOT THE SAME NEPTUNE

The trip to China transformed Heidi. There she saw gifted musicians and performers living in conditions that many Americans would consider unacceptable. Their crowded houses surrounded a courtyard, where an outdoor faucet in the middle provided their only running water. It made her realize what a great opportunity she had as a student at the Institute for Astronomy. Also, her new attitude and status as a minor celebrity transformed Heidi in the eyes of her fellow students and professors. Now nothing stood in the way of her passing when she took the quals for a second time. And that's just what she did.

Neptune, as it appears photographed by *Voyager 2* in August 1989 *(opposite)*. Heidi *(above)* visits Hawaii's Kilauea Volcano, one of the world's most active volcanoes.

~ Choosing Planetary Science

When Heidi first arrived in Hawaii, her research experience was mainly in planetary astronomy. But in 1984, after two years of advanced courses, she now had a broader outlook on the field. She had taken courses from professors who were experts on topics that Heidi had only studied in books. Now she could select a research project on bodies that orbit the Sun, stars, galaxies, or even the evolution of the universe. Which would she choose?

As usual, Heidi looked for a challenge that would help her grow as a professional and that matched her talents. She found both in the far reaches of the solar system. In spite of all the new areas she had been exposed to, she stuck with planetary astronomy after all. For Heidi, planets offered the chance to feel like an explorer. So little was known about Uranus and Neptune that she could be a pioneer.

For the last three years of graduate school, she continued studying the giant planets Uranus and Neptune. Imaging those distant worlds was a tough project. After all, Uranus is at least 1.6 billion miles from Earth, and Neptune is at least 2.38 billion miles away. But some of the world's expert observers were on the faculty of the Institute for Astronomy and could help Heidi with her work. The University of Hawaii also had an instrument that would make the tough job possible, an 88-inch telescope near the summit of Mauna Kea, the tallest mountain in the state.

Astronomical imaging is not the same thing as taking a photograph. Besides collecting light and producing a picture of sorts, imaging involves interpreting data and drawing conclusions about the object being observed. Through her images, Heidi could learn what the clouds on these distant worlds are made of or how they change with time.

Heidi and her mom *(above)* hike Kilauea in 1984. This methane band image of Neptune *(below left)* was obtained by the University of Hawaii 88-inch telescope at Mauna Kea *(below right)*. The telescope is used mainly by UH faculty and students.

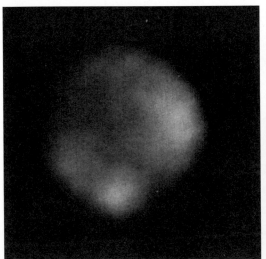

What Is Astronomical Imaging?

Instead of taking pictures in regular visible light, astronomers often use a filter to separate out one particular wavelength, or color. By looking at wavelengths separately, they can learn about the composition of an object. Heidi most often does imaging using a red filter to separate out near-infrared waves, which are longer in wavelength than unaided human eyes can detect.

Why does Heidi use a red filter to image the ice giants? Both Uranus and Neptune have methane gas in their atmospheres. Methane molecules absorb red light. This gives scientists a great way to better detect the location, altitude, and movement of the planets' clouds. Here's how: As sunlight travels deeper into the atmosphere, more of the red light from

its spectrum gets absorbed by methane molecules. If we use a red filter that blocks out all other wavelengths of light, the deeper we look into the atmosphere, the darker it will appear. Since clouds reflect the sunlight that hits them, the higher they are in the atmosphere, the brighter they will appear. It's because the methane molecules haven't absorbed all of the red light. But the lower the clouds are, the more dark or invisible they will appear through a red filter. The pictures of Uranus below show this. On the left is what Uranus would look like if we were in a spaceship heading for the planet—it's pretty featureless. But through a red filter (*right*), we can see a banded structure with the brighter areas showing clouds higher up in the atmosphere.

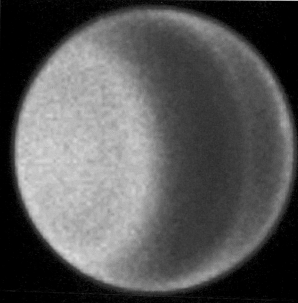

Since Heidi's work involved data, it drew on her most valuable skills. She instinctively understood what it took to create useful images, and she had a gift for interpreting them. The images of the planets allowed Heidi to understand the motions of the planets' clouds, the speeds of the winds, and the evolution of clouds over time. Almost nothing was known about Uranus and Neptune then, so every picture told scientists something new and unexpected.

Heidi's work led to another great opportunity. In 1986, *Voyager* 2, an unmanned space probe launched by NASA in 1977 to explore the giant planets, had reached Uranus. Heidi was one of several young scientists invited to NASA's Jet Propulsion Laboratory (JPL) in California to witness the space probe's close-up pictures as they came in.

Voyager 2's images revealed fascinating information about the moons, rings, and magnetic field of Uranus. For the astronomers studying those aspects of the planet, the mission provided plenty

On board each Voyager spacecraft is a 12-inch gold-plated copper disk with sounds and images from Earth and greetings in 55 languages—just in case anyone out there is listening.

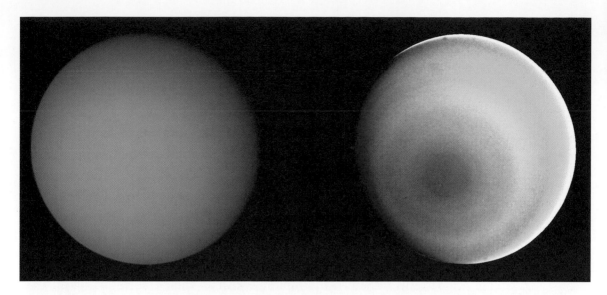

of new data. Unfortunately for Heidi, the images of the planet's atmosphere (her area of interest) weren't as interesting as she had hoped they would be. The planet had few clouds, and Heidi and the other scientists thought its atmosphere was bland. Fortunately, the *Voyager 2* mission had one more planet to explore. As the space probe swung around Uranus, the thrusters adjusted its course and sent it on a path toward Neptune. That planet might have a more interesting atmosphere.

It was very difficult to see much detail on Neptune from Earth. Heidi was eager to see pictures of that planet from *Voyager 2*'s closer viewpoint, but she would have to be patient. The space probe's arrival there was set for 1989, three years from now. Surely it would turn up a surprise or two.

Taken by *Voyager 2* on January 17, 1986, the true-color picture of Uranus *(left)* shows how someone aboard the spacecraft might view the planet. The false-color picture *(right)* shows an enhanced version of the true-color image to reveal its banded structure.

~ *Focus on Neptune*

In the meantime, Heidi returned to Hawaii. She was excited to have seen the *Voyager 2* team's work up close and was determined to be more than a mere observer next time. She wanted to actually work with the imaging team when *Voyager 2* approached Neptune. To prepare herself, she decided to focus on that planet for the rest of her research in Hawaii.

It didn't take long for her imaging results to stir up a little controversy and a lot of interest. Other astronomers had been producing images of Neptune's atmosphere since the late 1970s.

Now in 1986, using the same telescope and the same techniques she had used so successfully with Uranus, Heidi produced new images of Neptune's atmosphere.

To her astonishment, it was not the same Neptune! Instead of many clouds in both the northern and southern hemispheres, Heidi's images showed a single bright cloud—what would later be named the Bright Companion. Other scientists had their doubts, especially since Heidi had not yet established her reputation. Their first reaction was, "Well, you must be doing it wrong."

They had good reason for that line of thinking. The atmospheric patterns of Saturn and Jupiter—with all their colorful bands and spots—hadn't changed significantly for decades. And Uranus's atmosphere was plain and uninteresting. The four giant planets were close cousins, so why should Neptune's atmosphere be so changeable when the cloud patterns of the other three were stable?

Heidi insisted that her results were right. "I'm taking a picture. What could I be doing wrong?" she argued. It wasn't quite as simple as that, but eventually Heidi persuaded her critics that the atmospheric changes were real. But that wasn't the only scientific wrangling she got into over Neptune that year. Another controversy erupted when Heidi presented her results at a meeting of the Division for Planetary Sciences of the American Astronomical Society in Paris. This time the disagreement was over how fast the planet was rotating.

Neptune, Uranus, Saturn, and Jupiter *(left to right)* are plucked from their orbits and set in a dramatic montage.

If you want to image moving clouds on a spinning planet, you need to know when they will be back in your field of view. It was important for the *Voyager* 2 imaging team to use the correct rotation rate for the clouds, meaning the number of hours before a cloud comes into view again. When Heidi calculated the rate for the specific clouds they wanted to study, however, her rate was faster than the accepted rotation rate. Fortunately, it was her number that was used to determine the picture-taking sequence for *Voyager* 2's encounter with Neptune.

When Heidi finished her project and earned her doctorate in 1988, she received a postdoctoral position as a resident research associate at the Jet Propulsion Laboratory. Just as she had hoped, she would be part of the mission when *Voyager* 2 flew by Neptune. It was quite unusual for the Voyager team to invite an unknown young person to be part of such an important group, so Heidi was thrilled. But she wasn't surprised. She had set out to make herself the expert on Neptune, and that's what she was—and she was still only 29 years old.

Doing serious graduate work in science doesn't mean you have to be serious all the time.

Heidi's work

was giving atmospheric scientists

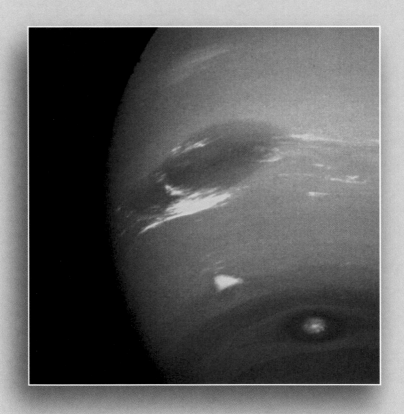

clues for **understanding**

Neptune's weather.

8

Neptune's Great Dark Spot

s *Voyager* 2 approached Neptune in August 1989, its
pictures produced a bigger surprise than anyone had
hoped. Until then, astronomers had only been able to
view Neptune from telescopes on Earth, about 2.38 billion miles
away from the planet. The view showed hints of dynamic clouds
but little detail. It was like trying to make out the features of a
dime from 10 miles away.

Now *Voyager* 2 was only 3,000 miles away from Neptune,
and the view was dramatically different. The difference was like
holding the dime in the palm of your hand. In the planet's southern
hemisphere was a huge hurricane-like storm that was soon called
the Great Dark Spot, which Heidi helped name. If a storm this
big were on Earth, it would cover the entire planet. Near the
Great Dark Spot, *Voyager* 2 measured winds blowing up to 1,500
miles per hour, the strongest measured on any planet. Fast-moving
clouds, such as a little white one dubbed "Scooter," zipped across
Neptune. The whole planet, in fact, was percolating with cloud
features and storms that changed rapidly as the spacecraft observed
the planet. All the activity in Neptune's atmosphere left Heidi and
the other astronomers incredulous.

This image of Neptune (*opposite*) was constructed from two Voyager images shot through green and clear filters. South of the Great Dark Spot is "Scooter," with Dark Spot 2 brewing farther below. Neptune's haze, found above the planet's clouds, is captured in false color (*above*).

~ The Cold Glow of Neptune

Heidi wished she could be with her colleagues at the Jet Propulsion Laboratory in California for the flyby. She wanted to watch with them as the data came in from Neptune, as she had done for the Uranus encounter. But she had a more important job to do for the mission back in Hawaii. While *Voyager 2* was sending back its photographs of the planet in visible light, Heidi was using the University of Hawaii's 88-inch telescope to get red filter images of the planet.

The clouds Heidi saw through the red filter were completely different from what *Voyager 2* saw. Heidi didn't see the Great Dark Spot. Instead she saw a very bright cloud nearby. As the Great Dark Spot rotated across the planet in the *Voyager 2* images,

From a range of 98,000 miles, *Voyager 2* records dramatic bright cloud streaks. Some of the clouds cast shadows on deeper clouds below.

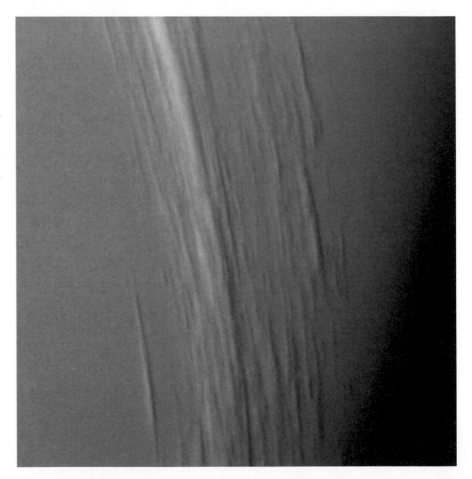

Roving Robotic Reporters

Unmanned probes like the Voyagers are launched into space to investigate our solar system and report back to Earth. These highly sophisticated robots follow a mapped-out route and are often in space for years at a time. The twin Voyager missions, *Voyagers 1* and *2*, were launched in 1977 and are still traveling to the farthest reaches of our solar system.

As the probe nears its target—referred to as a flyby—instruments on board go to work collecting data and sending it back to Earth. The Voyager probes each have 11 instruments on board, including two cameras.

The Voyager space probes have explored and expanded our knowledge of the four planetary giants: Jupiter, Saturn, Uranus, and Neptune. But their journey isn't over.

Their current interstellar mission, begun in 1990, will explore the outer boundary of the solar system in search of the heliopause. The heliopause is the region where the Sun's influence ends and inter-stellar (deep) space begins. Sometime before 2008, NASA expects the two Voyagers to cross an area known as the termination shock. This is where the solar winds slow from 1 million miles per hour to about 250,000 miles per hour. It will take another 10 to 20 years for the probes to reach the heliopause.

Voyager 2 is the only spacecraft to have visited Uranus and Neptune. *Voyager 1*, traveling 38,000 mph, reached an incredible 90 astro-nomical units (AU) from the Sun on November 5, 2003. (The distance from Earth to the Sun is 1 AU; 90 AU equals about 8.4 billion miles!)

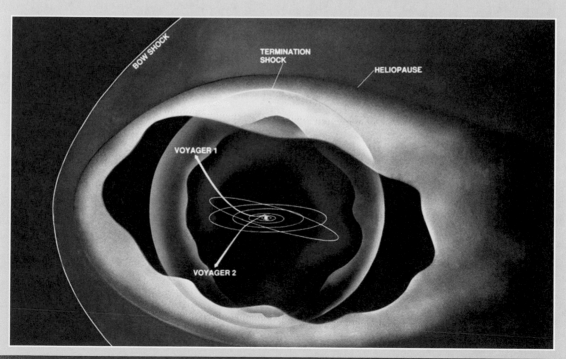

Taken 18 hours apart and over two rotations of Neptune, these images of the Great Dark Spot *(right)* show just how rapidly the planet's clouds can change. JPL scientists *(below)* examine data from *Voyager 2*'s flyby of Neptune in August 1989.

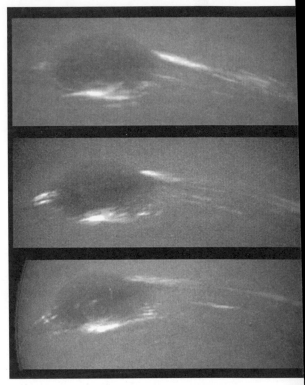

Heidi's cloud followed it in her red filter images. The two were related! Heidi knew exactly what her images were telling her about Neptune, and she knew immediately that the results were important. The Great Dark Spot was only part of the scientific story. Heidi was once again seeing the Bright Companion, and it meant that many of the other bright features she and her colleagues had seen in the past could also have been bright companions to great dark spots, even though they hadn't been able to see those spots.

The Great Dark Spot was acting like a mountain would on Earth. Mountains cause the air to flow around and over them,

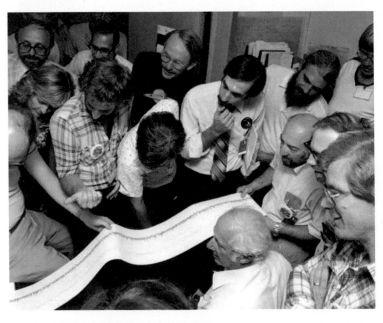

which in turn causes a change in air temperature and pressure. That change often makes clouds form near the mountain. Similarly, the Great Dark Spot was causing Neptune's atmosphere to change temperature and pressure, forming the Bright Companion. Heidi's work was giving atmospheric scientists clues for understanding Neptune's weather. This was important because the more we know about the physical and chemical properties of atmospheres of distant planets and how they function, the better we can understand our own atmosphere.

As excited as Heidi and her colleagues were about her results, it didn't make up for her disappointment at not being at the Jet Propulsion Laboratory. The discovery of the Great Dark Spot and other features on Neptune as *Voyager 2* approached that planet had created a lot of public interest in the flyby. Television news and documentary film crews swarmed around the scientists in California as the data came in, and ordinary people all over the world got to share the thrill of discovery beamed to Earth from more than 2 billion miles away.

> After the *Voyager 2* flyby of Neptune was over, the world had a different picture of that distant blue world, and scientists were bubbling with new questions about it.

Unfortunately, because Heidi was probably the only person in the world who had the experience to do the methane band imaging, she didn't get to appear in any of the documentaries. She wanted to be a part of all the media attention, but because she was on an isolated mountaintop thousands of miles away from the action, she missed out on all the hoopla.

After the *Voyager 2* flyby of Neptune was over, the world had a different picture of that distant blue world, and scientists were bubbling with new questions about it. How long do those cloud features last? What causes them? Why should Neptune have a lot of cloud features and Uranus have almost none?

Fortunately, they wouldn't have to wait for another space probe flyby to get some answers. The Hubble Space Telescope would soon be launched. This Earth-orbiting instrument collects and focuses light with a giant mirror to form images that it sends back to scientists as radio signals. It can see 10 times better than the most powerful telescope on Earth. Scientists—Heidi included— were already jostling to be first to turn its planetary cameras toward Neptune.

~ Hubble Trouble

Less than a year later, on April 24, 1990, the space shuttle *Discovery* lifted off with the Hubble Space Telescope in its cargo hold. Once deployed, the telescope would orbit Earth, capturing images from above its atmosphere. Astronomers all over the world had already submitted proposals to observe their favorite celestial objects without having to worry about Earth's atmosphere interfering with their view. Because of *Voyager* 2's discoveries, Neptune and the Great Dark Spot were high on the list of what scientists wanted to look at.

Space shuttle *Discovery* rises above the Kennedy Space Center liftoff pad with the Hubble Space Telescope tucked away in its cargo bay.

Then came the bad news. Hubble's vision was distorted. The telescope's primary mirror should have created crisp, clear images, but somehow the mirror had been ground to the wrong shape. The flaw was tiny, measuring only 2 microns, which is about one-fiftieth of the width of a human hair. But that was enough to prevent the lens from focusing properly. The bad news was that the flawed primary mirror could not be replaced with a new one. However, the good news was that it would be possible to correct the problem with new optics. Scientists wrote an "eyeglass" prescription for the orbiting observatory. Unfortunately, it would take three years for the repair to be made.

Heidi takes a break from her hectic data-filled day.

After Heidi's postdoctoral position at the Jet Propulsion Laboratory ended in 1990, she came to another turning point. She needed to find something else to work on. She was sure she could find a good job, but it was important to choose the right one.

Wherever that job would be, Heidi was eager to use the Hubble to study the giant planets. She could use the delay caused by the space telescope's blurred vision to her advantage. Three more years of work would cement her reputation. So Heidi looked for a job where she could take the lead on some important projects to image Neptune and the other giant planets. She found it at a place she had wanted to leave forever only eight years earlier—MIT.

Heidi *was used*
to sending out *proposals*
for her job,

but on Christmas she found
herself on the receiving end.

Unexpected Developments

9

Heidi's move back to Cambridge had an unexpected benefit—romance. Long before there was the World Wide Web, Heidi had actively participated in an e-mail discussion group about the Grateful Dead. Even after she returned to MIT, she continued to use her Hawaii e-mail address to post messages to the group because she hadn't yet figured out how to access it from her MIT account. When she saw that another participant, Tim Dalton, was posting from MIT, she decided to send him an e-mail asking how to make the changeover.

As they talked back and forth, first in e-mail and then in person, Heidi and Tim discovered that they had more in common than an interest in the Dead. Like Heidi, Tim was serious about his work (chemical engineering), but he recognized the value of having a life outside of his job. Most important, Tim was not intimidated by Heidi's success, and he and Heidi respected each other's goals. Their friendship quickly blossomed, and they became a couple.

Heidi and Tim *(opposite)* travel to Albuquerque, New Mexico, to attend Hazen's wedding. A partial view *(above)* of a Neptune image that was captured by Hubble's Wide Field Planetary Camera 2.

~ Proposals, Proposals

Endeavour astronauts Story Musgrave and Jeffrey Hoffman wrap up their final spacewalk of the Hubble Telescope repair mission, with outstanding "before and after" results *(far right)*.

At MIT, Heidi was a research scientist in the Department of Earth, Atmospheric and Planetary Sciences. One of her most important jobs was writing proposals to request telescope time and funding for observations she and her colleagues wanted to carry out. She submitted a proposal to use the Hubble Space Telescope to study Neptune. She wanted to follow up on the *Voyager* 2 images to get the best pictures she could of that planet. And she wanted to see how Neptune looked five years later. In December 1993, her proposal was accepted. That same month, she joined astronomers around the world in cheering the crew of the space shuttle *Endeavour*

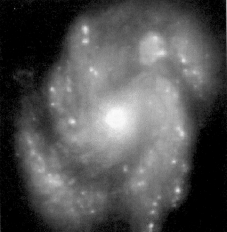

as they installed corrective optics on the Hubble. She would now be able to see the Great Dark Spot again.

Heidi was used to sending out proposals for her job, but on Christmas she found herself on the receiving end. Not long after her Hubble proposal was accepted, Tim asked her to marry him and presented her with an engagement ring. She was ready for another adventure, and marrying Tim seemed like a good thing to do. So she said "yes."

~ Vanished!

It was now early 1994 and Heidi eagerly looked toward Neptune. When her turn came to use Hubble, she studied the images with great anticipation. The Great Dark Spot should have been easy to pick out. But she couldn't believe what she was seeing—or rather

The changing Neptunian weather system is full of surprises for Heidi and other scientists—but the unexpected often leads to increased scientific knowledge.

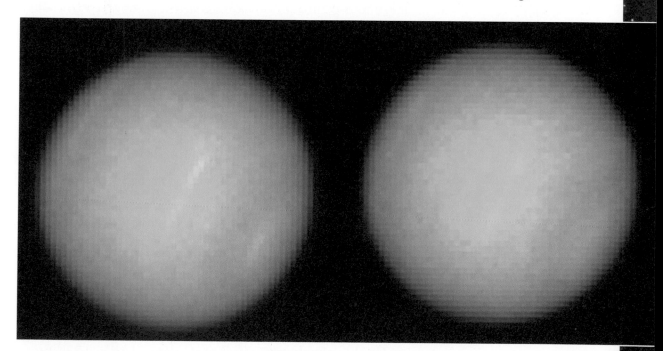

not seeing. The Great Dark Spot was gone—vanished! Her favorite planet had delivered another surprise.

Neptune's atmosphere was more changeable than scientists had imagined. Figuring out why would take careful observation.

Heidi knew right away that she and other scientists would have to gather data for years to understand what was happening with the weather on that distant giant world. It's not just the infrequency of data collection that makes the process so long. Neptune takes an extraordinarily long time to orbit the Sun—the equivalent of 165 Earth years! To observe only one "month" on Neptune, an astronomer would need decades of Earth time.

She also knew that most changes wouldn't be as obvious as the appearance and disappearance of the Great Dark Spot. Smaller Neptunian weather systems must also be developing and disappearing, but they would be much harder to detect. To find them she would have to hunt for clues hidden in a mass of detailed measurements. And she didn't even know what those measurements would reveal. Were they chemical changes? Cloud movements? Or something else? It would be like looking for needles in haystacks, but she knew she would have fun burrowing into the pile of data and organizing it, making charts and graphs as she did in her creative Monopoly games years earlier in Clarks Summit.

~ The Great Comet Crash

The disappearance of Neptune's Great Dark Spot was a big story. But another solar system event in 1994 had even greater impact. That one would be a turning point in Heidi's career because it transformed her into a celebrity.

It began on March 25, 1993, when the International Astronomical Union issued a circular—a breaking-news report—about an unusual discovery. Carolyn Shoemaker, Eugene Shoemaker, and David Levy had found their ninth comet. Almost immediately, Shoemaker-Levy 9 became one of the most famous comets in history. It was unlike any other comet ever seen.

Comets are big chunks of ice and dust, sort of like giant dirty snowballs. Most comets appear as a hazy ball of light with a long tail extending away from the Sun. This one, however, looked like a bar with several tails. Carolyn Shoemaker called it "a squashed comet."

If Heidi saw the circular announcing the discovery of periodic comet Shoemaker-Levy 9, she doesn't recall. She doesn't even remember reacting to later circulars announcing that the comet was orbiting Jupiter, which accounted for its odd appearance. On its most recent orbit, it had come so close to Jupiter that the giant planet's powerful gravity broke it

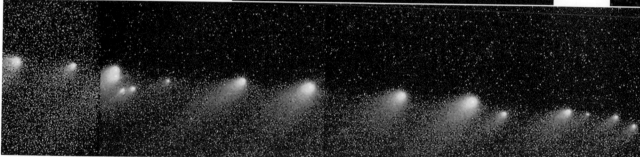

into fragments. The breakup released enough dust to make it visible from Earth. When large Earth-based telescopes and Hubble focused on the comet, many observers described its appearance as "a string of pearls."

Heidi saw newspaper accounts and scientific articles about the comet but didn't pay much attention to them. It didn't seem to have anything to do with her work—even when a May 1993 circular reported details of the comet's orbit. Heidi heard about it from a graduate student, who came to her office with more enthusiasm than she thought the news deserved.

"They found this comet," the student announced.

"Yeah?" said Heidi.

"It's going to hit Jupiter!" he continued. "Isn't that exciting?"

"Well, you know Jupiter is a really big planet," Heidi replied. "Comets are really small. I don't think much of anything's going to happen when this comet hits Jupiter."

Comet West (top) with its magnificent dust and ion tails dazzles onlookers in 1976. All aboard the comet train! A mosaic image of 20 Shoemaker-Levy 9 comet fragments (above) as seen by Hubble.

She finally appreciated the importance of the event when colleague Tim Dowling approached her about writing a proposal to image the impact with Hubble. The comet was due to crash into Jupiter in the summer of 1994. Tim had created a mathematical model—a set of equations—to predict what would happen when pieces of the comet hit Jupiter. His model showed huge ripples in Jupiter's atmosphere. It looked very dramatic, but Heidi was doubtful.

"Give me a break," she told him. "This is not really going to happen."

"No, no!" Tim persisted. "I really think we ought to look at this with the Hubble Space Telescope."

"Fine," said Heidi. "I don't believe in it, but what the heck." NASA was already putting money and Hubble time into studying the comet impact, and the bandwagon was rolling. Even though Heidi had her doubts about what would happen on Jupiter, Tim's model looked interesting. She decided it was worth her time to write a proposal. In the proposal, she requested permission to record the comet's collision with Jupiter's atmosphere using Hubble at visible wavelengths of light.

> This was a terrific opportunity, but as the least experienced and youngest among the five, she wondered if she would be able to hold her own.

In January of 1994, six months before the event that came to be called "the Great Comet Crash," Heidi got a phone call from NASA. Her proposal had been accepted! But there was more. "We've reviewed all these proposals," the caller told her, "and we've decided that we are accepting five projects." When she heard the names of the other four scientists who also had Hubble imaging proposals selected, Heidi's mind started racing. Every one of them was outstanding.

This was a terrific opportunity, but as the least experienced and youngest among the five, she wondered if she would be able to hold her own. So she couldn't believe what she heard next: "We want to combine them into one proposal, with you as the principal investigator." In science the principal investigator is the person who oversees the entire project.

Heidi was astounded but as usual kept her cool. She wasn't sure that she was ready to lead a team of top-notch astronomers. She couldn't say "no," but she wasn't ready to say "yes" either. "What a great idea," she said. "Let me think about that and get back to you." Then she hung up.

She reviewed her options. Her proposal was selected along with those of four top-ranked Jupiter atmosphere experts. She had looked at Jupiter's atmosphere but never published a paper about it and was not an expert on it by any means. And now they were asking her to be the principal investigator. *This is insane,* she told herself. *But of course, I have to say "yes."*

An **event** like this
had never been seen before.

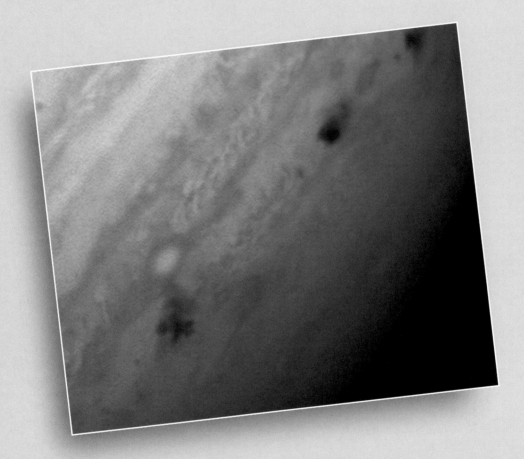

No one—including Heidi—knew
exactly what to expect.

10

JUPITER'S BLACK EYE

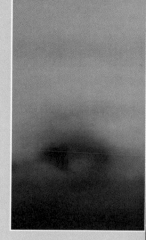

W ith her selection as principal investigator on a major Hubble Space Telescope imaging project, Heidi's career as an astronomer was taking off. She wasn't surprised by her success—she had worked hard for it—but she was about to be surprised by the direction the Great Comet Crash would take her.

By the time the contracts were all signed, the members of the team had barely five months to prepare to observe the impacts of at least 20 comet fragments over a week in mid-July of 1994.

Dark comet impact sites on Jupiter are clearly visible in this color composite *(opposite)*. Hubble also took ultraviolet light images of the Great Comet Crash *(above)*.

~ Planning for the Crash

An event like this had never been seen before. No one—including Heidi—knew exactly what to expect. For all anyone knew, the event could turn out to be a complete dud. They weren't even sure what they should be looking for. And if that wasn't stressful enough, the Great Comet Crash would be a once-in-a-lifetime occurrence. That meant it was crucial to get as much good data as possible. No mistakes. Heidi and her colleagues would have to draw on all their experience to anticipate problems before they happened.

Heidi gathered whatever information she could find on Hubble's cameras and its orbit. She studied the available data on the trajectories (paths) of the comet fragments. She drew diagrams and made plans for how the observing would go. If Heidi had anything to do with it, the imaging would go off without a hitch.

~ Disaster (Almost) Strikes

The day before the first fragment of the comet was due to plow into Jupiter the team felt about as ready as they ever would. With months of careful preparation behind them, all that was left was to let the cameras on the Hubble Space Telescope run.

So when Heidi and her colleagues stared into the computer screen at the Space Telescope Science Institute (STScI) in Baltimore, Maryland, they expected to see the last images of Jupiter before the impact. Instead, what she saw made her heart stop. The images of Jupiter that were taken through a red filter were overexposed! If the commands for Hubble's planetary camera were not adjusted, there would be no red filter images to analyze—and no color pictures of Jupiter.

In the audience were journalists from major news services, magazines, and local newspapers from across the United States and around the world.

Heidi looked at the big clock on the wall. In five minutes a week's worth of new commands would be sent to the space telescope. There wasn't a moment to spare. She and her team members rushed to the people in charge and asked them to change the exposure times for the red filter images.

"But we've never done such a thing," one of the Hubble people responded.

"You don't understand. We have to fix this," Heidi insisted.

"We can't. It's impossible. We have to follow normal procedures."

"We don't have to change all the commands. We only have to figure out and change one number," Heidi assured them. The clock was ticking and they were running out of time.

"Can you figure out the number in five minutes?"

"We'll have to."

They delayed the transmission—by a scant five minutes—so that Heidi's team could recalculate the red filter exposure times.

~ Heidi Takes the Stage

The next day—July 16, 1994—a buzz filled the briefing room at the Space Telescope Science Institute. On the stage were Eugene Shoemaker, Carolyn Shoemaker, and David Levy, the team that less than 16 months earlier had announced the discovery of the unusual object known as Comet Shoemaker-Levy 9. In the audience were journalists from major news services, magazines, and local newspapers from across the United States and around the world.

The first fragment of the comet—racing at speeds of about 130,000 miles per hour—had already plunged into Jupiter's cloud tops. Before the news conference was over, the reporters hoped to see Hubble's first images of the event.

One floor below the briefing room, Heidi and her imaging team were gathered around a computer screen waiting impatiently.

Heidi and astronomers at the STScI huddle together and play the data waiting game, something many scientists are used to doing.

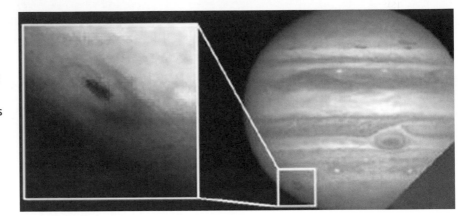

In this close-up by Hubble, the impact site of fragment-A is seen as a dark streak and a crescent-shaped feature measuring several thousand miles across.

The fragment had struck the planet on the far side of Jupiter, just out of Hubble's view, but the planet would soon rotate far enough for the impact site to be seen. The next time Hubble's orbit around Earth brought it to the point where it could view Jupiter—90 minutes later—the space telescope would transmit an image of the crash site.

Upstairs, Eugene and Carolyn Shoemaker each gave brief reports on some early observations of Jupiter. They then turned the microphone over to David Levy, a noted science writer and amateur astronomer who had become the Shoemakers' trusted coworker in the observatory.

Levy began his presentation by referring to the 25th anniversary of the first human Moon landing, only four days away. That was one of those "once-in-a-generation moments when science stops, turns on its ear, and gives us something really, really special." The impact of Comet Shoemaker-Levy 9 was also going to be such a moment, showing people everywhere that science is full of surprises. He also emphasized the significance of the impact. "It's likely that cometary impacts provided the water that's on the Earth right now," he said of long-ago comets that pummeled our own planet. Levy suggested that everyone should go out and observe Jupiter on this historic week. "If you only have a small telescope, you should use it." But he also warned, "You're probably not going to see Jupiter show major changes from this impact."

Levy was wrong, and Heidi and her imaging team already knew it. They had been watching, incredulous, as evidence of the impact appeared on their screen. Their display showed a bright plume—

an arch-shaped column—of gas rising more than 600 miles above the edge of Jupiter from an impact site just on its far side. The Hubble Space Telescope had come through with flying colors. The images were full of sharp, rich detail that would give scientists loads of new data they could use to learn more about the atmosphere of Jupiter—not to mention comets.

The team burst into cheers and celebrated Hubble's feat with a bottle of champagne. Now all they had to do was wait for Hubble's next orbit to actually view the impact site.

Ninety minutes later the next Hubble image clearly showed the impact site. It was huge. So big, in fact, that it would be visible from Earth even to backyard observers with only small telescopes, watching from 365 million miles away. This was not just some obscure discovery to get astronomers excited. This was everybody's event.

Heidi printed a copy on the laser printer and headed toward the briefing room. Gene Shoemaker was answering a reporter's question about the possibility of a "great fizzle" when a beaming Heidi approached.

This computer-generated representation shows Comet Shoemaker-Levy 9 as it passes above Jupiter's south pole on its way to final impact.

She showed the image to Gene, who offered her his seat. It wasn't hard for the reporters to tell she had big news. Her sparkling, expressive eyes and broad smile said it all. "Gene Shoemaker said he would be personally astonished if we didn't see something," she began. "Well, he is not going to be astonished. We actually saw some amazing things! We were able to see

a plume on the edge of the planet. And I'll remind you that this is fragment A, the first one—not the brightest one. So we're going to have a really exciting week."

Heidi displayed the image for the cameras, like a proud parent showing off a photo of a newborn baby. "See that bright streak?" she said, her voice rising with excitement. "And around the edge of the streak, there's some other stuff," she continued, repeating the word "stuff" to laughter all around, including her own. "Tell you more about the stuff tonight."

The reporters were captivated. Heidi's energy and unbounded excitement were contagious. They were not used to such an animated performance and vivid language from a scientist. As Heidi strode off, they erupted into applause.

Back in the observing station, Heidi had less than two hours to prepare for a news conference with the other Shoemaker-Levy 9 principal investigators.

~ The Guy on the Street

At 10 P.M., Heidi and the four Hubble principal investigators of other comet-crash projects took their places at the briefing table. Reporters asked a number of scientific questions, and each scientist answered carefully, to make sure the press got the details straight.

As informative as the other scientists were, it was clear to everyone that Heidi was an altogether different kind of scientist. Whenever she answered a question, her body was as animated as her voice. And she spoke in terms that everybody could relate to. There was no mistaking her excitement for her work.

Then a reporter from BBC-TV asked about people who normally don't think about astronomy. "What's in it for the guy on the street?" he asked. One of the other principal investigators spoke up, getting a laugh for declaring they should be glad they don't live on Jupiter. Then he turned serious and talked about the importance of science and technology and the historic rarity of the impact.

When he finished, there was a slight pause. Then Heidi spoke up. She was careful not to sound critical, but she knew her

Against a brilliant splash of extraterrestrial color, Heidi takes her place onstage with David Levy *(left)* and the Shoemakers, as they prepare to recount the exciting events.

colleague had missed the point. The "guy on the street" represented men and women who ordinarily weren't interested in science but who knew enough to see the importance of this story.

"There's a little bit more to it than just the science," she said. "Of course, we're interested in science here because that's what we do. But it's a fascinating thing. There are things whizzing around the solar system, smashing into other things with huge explosions, and that's just really incredible to think about." Then she got a little more serious. "We don't often think about the universe out there. We just sort of look at the sky and stars up there and—big deal. But if we really take a step back, it's a dynamic universe. And this is just a key example of some of the energetics that go on."

That wonder and awe, together with her plainspoken and humorous answers, made Heidi a favorite of both the reporters and the everyday people who were as interested in the Great Comet Crash in their own way as the scientists were in theirs.

At one news briefing after another that week, Heidi told the story of the impacts in colorful terms, speaking of a bruised planet with a series of black eyes. She was the young, enthusiastic human face of the Great Comet Crash, and everyone loved it.

It was time to get back to work
researching Uranus and Neptune

and to **reconnect** with the rest
of her life.

ASTRONOMY'S NEWEST STAR

11

After the excitement of the Great Comet Crash, Heidi returned to MIT as a celebrity. The July 27, 1994, issue of the *Chronicle of Higher Education*, the national weekly newspaper for college professors and administrators, raved about her ability to reach the public. The headline called Heidi "The Comet Drama's Biggest Hit" and "astronomy's newest star." The article itself gushed: "The energetic 34-year-old research scientist from the Massachusetts Institute of Technology demonstrated to the world that some astronomers know how to have fun."

But Heidi knew that the fun in science is rarely about big surprises like bruised planets and a disappearing Great Dark Spot. Most scientific work is made up of lots of little details, with occasional small surprises to make it interesting. To be on hand for the surprises, you have to live the details and enjoy them.

Heidi understood that everyday life is that way, too. After being at the center of one of astronomy's most eventful weeks ever, she was ready to go back to the day-to-day details of working and living in Cambridge, Massachusetts. It was time to get back to work researching Uranus and Neptune and to reconnect with the rest of her life—especially Tim.

A clearly happy Heidi *(opposite and above)* enjoys her well-earned success but, not surprisingly, she is already planning her next move.

~ Marriage, Children, and a Move

With her career well established, Heidi was ready to think about marriage and a family. But she knew it would be a big adjustment. She and Tim decided to enjoy being engaged and not rush the wedding. They had been dating for three years before Tim gave Heidi the engagement ring. It would be nearly another three years before they married in 1996.

After Heidi and Tim's wedding, it was natural for them to consider starting a family. Heidi's thoughts about having children were similar to her reaction to Tim's proposal. It was not a goal in her life, but she was open to the idea. Her career was fine, so she decided, *Well, if it's going to happen, it's going to happen.* It happened, not just once but three times.

Heidi's first pregnancy was so easy that she worked until the day before she gave birth. Her daughter, Beatrix, was born in 1997, 10 days overdue.

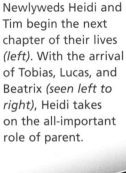

Newlyweds Heidi and Tim begin the next chapter of their lives *(left)*. With the arrival of Tobias, Lucas, and Beatrix *(seen left to right)*, Heidi takes on the all-important role of parent.

The doctors had talked about Heidi's "advanced maternal age" of 37, but she had no problems at all.

Her second pregnancy was a bit harder, and there was an additional complication. Tim had taken a job with IBM in New York's Hudson Valley, and he and Heidi moved to western Connecticut. Heidi had a three-hour drive to MIT from there, but she had already cut her work to part-time and could do most of it from home. Still, at least once a week, she had to be in the office, and so on those days she spent six hours on the road—even late into the pregnancy. Soon after giving birth to son Tobias in 1999, Heidi went back to work at MIT, and back on the road to Cambridge.

As much as Heidi loved being a parent, it was hard work. She soon realized that going to MIT every week was too stressful, so she decided to look for a new job. Heidi was delighted when the Space Science Institute (SSI) in Boulder, Colorado, offered her a position as a senior research scientist and arranged for her to work from home every day.

Heidi's third pregnancy at age 41 was relatively uneventful. However, it was certainly harder than the first two, and there were days when she felt miserable.

"You're getting too old for this kind of thing," her doctor told her. "You've got kids at home, you work full time, you travel extensively, and you're tired. That's why you feel so rotten." When she gave birth to her son Lucas in 2001, she decided that three children were enough.

~ Family Life

Today, Heidi doesn't hesitate to tell people how grateful she is. She and Tim have a strong marriage and bright, happy, healthy children. But she knows that good fortune alone is not enough. "It's just like everything else in my life," she will tell you. "This was the opportunity that presented itself, and we did everything we could to make sure it would be a good experience. And it has been."

Heidi and Tim take their roles as mom and dad seriously but with an easy good humor. One look at their household calendar shows that they schedule their time carefully and make sure that no detail escapes them. They both travel frequently for work, but they always make sure that one of them is home. Sometimes they call on Heidi's mother, called Graga by the children, to give them a hand.

On work days when both Heidi and Tim are home, Tim leaves early in the morning, and Heidi takes care of breakfast, reading a book to each child, and getting each one to school or a day care program, which she and Tim have chosen carefully. Then she gets down to being an astronomer in her home office until dinner. Tim is usually able to leave work by 4 P.M., so he picks up the children, brings them home, and cooks supper, a task he enjoys.

Heidi takes a break in her home office. Balancing family life and work is a priority for her and Tim.

Supper is family time, and then, after some playing and reading, the children go to bed. Tim also has a small office off the family room, so he and Heidi frequently spend part of the evening working at their computers taking care of leftover business.

~ Asteroid Hammel and Other Honors

Besides getting married and having children, Heidi's life after the Great Comet Crash has changed in ways she could never have anticipated. She is so well known that she is always fielding more invitations to speak about astronomy than she can handle.

Heidi has also received a steady stream of honors and awards for both her scientific work and her ability to share science with the public. On her office walls, interspersed with her children's artwork, are plaques and certificates from NASA, women's organizations, professional organizations, the famed San Francisco Exploratorium, and *Discover* magazine, which named her one of the 50 most important women scientists of 2002. She also proudly displays the 2002 Carl Sagan Medal of Excellence in Public Communication in Planetary Science, named after the famous astronomer and best-selling author.

Lastly, if you have a telescope and know when and where to look at the sky, you may spot an asteroid named 3530 Hammel. In 1996 this asteroid was renamed in Heidi's honor.

Heidi takes pride in these accolades, but she rarely mentions them when she speaks. She'd rather talk about science and the good things it has brought to her life.

And even though she loves to tell people about science, Heidi is happiest when she is observing. The universe is full of opportunities, and every night it offers the possibility of discovering something that no one else has ever seen.

That possibility is on Heidi's mind as she tries to sleep in the scientists' dormitory on Mauna Kea on a Sunday night in August 2003.

Heidi won the Spirit of American Women Award in 1996 for her many contributions to educational outreach, including working with young students.

A single night of **observing**
often produces enough
information to keep Heidi

and other **astronomers**
busy for months, sometimes even years.

12

DREAMS AND DISCOVERY

For Heidi, working at a modern astronomical observatory is like attending a Grateful Dead concert. Even when she knows every song, no two performances are exactly alike. Sitting in the audience, she can only observe, not control, events onstage. Often, too much is going on for her to catch everything.

The IRTF Telescope on Mauna Kea, Hawaii, is once again open to the night sky and to discovery. Even a partial glimpse *(above)* of the Milky Way can be awe inspiring.

Fortunately, concert performances can be recorded. Heidi's collection of Grateful Dead recordings includes many live concerts, some of which she attended. Every time she listens to them she discovers something new and surprising, often in places that had seemed ordinary before.

At a concert, Heidi concentrates on the performance. But for the next three nights at NASA's Infrared Telescope Facility, she will take a different approach: to get as much data as she can and figure out what it means later. Because her telescope time is so limited— as it is with all astronomers—she needs to make every second count. She can't spend her telescope time thinking about the meaning of certain data. Her only task is to record the information. A single night of observing often produces enough information to keep Heidi and other astronomers busy for months, sometimes even years.

That's true most of the time, but Heidi can already tell that this night, Monday, August 18, 2003, is likely to be different. She has been observing at Mauna Kea long enough to recognize a problem in tonight's spectacular sunset. Joined by the other members of her observing team, she gazes at the nearby twin domes of the Keck Observatory, where she plans other observations of Uranus and Neptune later in the year. Far below is a gray-white sea of cottony clouds. The Sun, like a giant red ball, sinks slowly into that layer—and then suddenly it disappears. The trouble looms above, where a few wispy high clouds contribute purple streaks to

The twin Keck Telescopes of Mauna Kea *(top)* are silhouetted against the sky. BASS team members Brad, David, Ray, Heidi, Fred (the author), and Daryl are in telescope heaven.

the spectacle. They are breathtaking in their beauty, but Heidi knows they are bad news. High clouds mean high humidity at the summit. She knows that telescope operator Paul Sears will not be allowed to open the dome until the humidity drops below 90 percent and stays there for 20 minutes.

Normally, Heidi works with red filters, gathering data from Uranus and Neptune. But tonight she will be going even farther into the red than the human eye can see. Known as infrared, this method measures a planet's heat rather than reflected sunlight. She's eager to get started. But waiting for the skies to clear is part of the astronomy game. There's nothing for her to do but hope the high clouds will dissipate and, in the meantime, enjoy the company of the other observers: Ray Russell, David Lynch, Daryl Kim, and Brad Perry.

Heidi looks at the clock. Uranus is rising, but the dome remains closed because of the humidity. She has been through

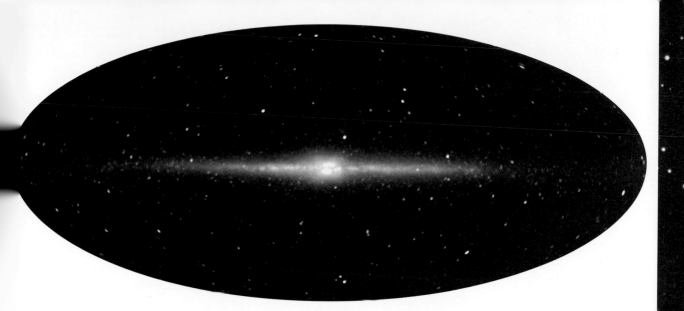

observation delays many times, but they are always frustrating. She walks outside and looks at the spectacular Milky Way high above. Mars is glowing bright red in the East, and Uranus is in the same general direction. If her eyes were fully adapted to the dark, she might just be able to spot its blue-green disk, which is visible without a telescope only in the darkest places on Earth. The thin scattered clouds pose no visibility problem tonight—but they add enough moisture to the air to keep the dome closed. The hours tick by.

At last the humidity breaks. This time, the downward trend is not a tease. By 12:30 A.M. the dome is open and the team is ready to make its first measurements.

Tonight the team will be observing with the Broadband Array Spectrograph System (BASS). This instrument attaches to the telescope and divides the light into different wavelengths. But they need to test the system first to make sure it is operating properly. If all goes well, no engineering problems will crop up, and they will be able to collect scientific data.

The more light Heidi can collect from Uranus's faint glow, the better she will be able to determine its infrared spectrum. From the spectrum she can deduce what gases are present in its cloud tops. She already knows the atmosphere consists mostly of hydrogen and helium. Tonight she wants to take a closer look at how much ethane is in the atmosphere. From past imaging, astronomers know there's an extra brightness in the infrared that they can trace to the ethane in the atmosphere. But the exact color and temperature of

This shot of the Milky Way was taken in infrared light by the COBE (Cosmic Background Explorer) satellite, a powerful instrument that measures the remains of the Big Bang.

that gas change in ways they don't understand. Heidi would like to collect enough data to figure out what's happening in Uranus's atmosphere.

Since Ray is in charge of BASS engineering, he gets to go first tonight. In short order he has the information he needs. Dave is next. His main goal this week is to look at a star that seems to be in the process of forming a dust cloud. But by 9 P.M. that star is no longer visible. So he looks at another target object, mainly to test how BASS is working for his purposes.

~ *Target Uranus*

Balanced on ladders, Daryl *(left)* and Ray carefully troubleshoot BASS's circuitry.

Shortly after 2 A.M., it's Heidi's turn. Paul sets the telescope's controls to swing it toward Uranus. She confirms that the planet is properly positioned in the field of view, and they begin to record data. The signal, as expected, is weak, and Heidi is thinking that if all goes according to plan—

"Channel shift!" Daryl announces. Something has gone wrong with the data collection. To produce a spectrum, the electronics in BASS divide the light into ranges of wavelength that correspond to certain channels. One channel has gone bad, and it affects every other channel above it.

They reset the computer equipment and continue collecting data, but soon the same channel shift recurs.

First the weather, now the equipment! Will Heidi ever get the data from Uranus she traveled so far to capture?

Ray shuts down the equipment, and he and Daryl take a look at the BASS circuitry to see if they can find the problem. They don't find anything obvious, so they decide to start everything up again. It's now 3:45 A.M. Soon Uranus will be too low in the sky to produce a good signal. Disappointed, Heidi decides to use the remaining time to look at other objects, including Neptune. She gets some good data about the gases in Neptune's atmosphere, which she will analyze when she gets home.

~ A Martian Surprise

Now it's time for a little adventure. This is the opportunity for Brad to squeeze in his pet observations for this session. In a week Mars will reach its closest approach to Earth in 60,000 years, so this is a good time to study the "Red Planet" and especially its two tiny potato-shaped moons, Deimos [DEE-mos] and Phobos.

Brad goes to his laptop computer and runs a program that calculates the sky position of the planets and their moons. It's too late tonight to catch Phobos, but Deimos will soon be in its best observing position, about as far from Mars in the sky as it gets.

Heidi asks Paul to turn the telescope toward Mars. Once it is centered on the planet, Brad tells Paul how far and in what direction to shift it to find Deimos. A faint spot on the observers' video monitor indicates that they have zeroed in on their target.

At 4:50 A.M., after a long and often discouraging first night, the team is eager to see Deimos's spectrum. The first data appears on the computer screen, and Dave is quick to respond. "Is the computer set properly?" he wants to know. Deimos can't possibly be that bright. It's off the scale! Daryl makes a quick readjustment to the computer display, and the graph of Deimos's full spectrum comes into view.

This outstanding Hubble image of Mars reveals a huge cyclonic (swirling) storm just below the planet's polar ice cap.

Phobos, the innermost and larger moon, takes 7 hours to orbit Mars, while Deimos takes 30 hours. Both moons are named after attendants of the Roman god Mars. Phobos translates to "fear" and Deimos to "panic."

BASS's data matches the earlier scientific measurements, so they conclude that the instrument is probably operating properly. But instead of showing a gradual rise in intensity, the measurements increased far faster than expected before dropping back down to the anticipated level.

The most likely explanation for the data is that the surface of Deimos heats up like sand on a sunny beach. The infrared telescope, which senses heat, must have spotted this. Still, as the team's excitement grows, so does its scientific caution. Whenever a new result is as unexpected and dramatic as this one appears to be, the first thing to suspect is the measurement technique or equipment. The team's members also need to make sure that they aren't so hungry for good news that they leave their good judgment behind.

But it's late. The Sun will soon be rising, and they all need sleep. Paul shuts down the telescope, closes the dome, and leaves some notes for the day crew to attend to. He and the weary observation team head down to the scientists' dormitory for breakfast and a good "night's" sleep.

Phobos

Deimos

~ Following Clues and Sharing News

The surprise from the night before is enticing enough that the team adjusts its plan for the next two nights. The plans will include more observations of both Martian moons. By the time the team members wrap up their third night of observing, they have data from Deimos and Phobos, and the results indicate that both are hotter than expected.

To a scientist like Heidi, the results from Deimos that night are every bit as surprising as Neptune's Great Dark Spot and the

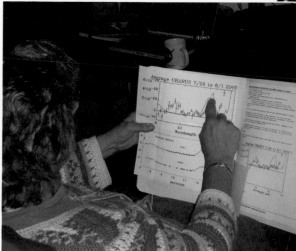

Great Comet Crash. Any time scientists arrive at new measurements that make them reconsider their interpretation of earlier measurements, it could lead to something new and important. And that may eventually have meaning for the general public.

Heidi has plenty of data from Neptune to keep her busy at work, but the observations of Uranus were a failure. Ray and Daryl suspect that a loose wire in BASS caused the channel shifts.

After three dusk-to-dawn sessions, the observers are all weary when they get back to the dormitory. This time Heidi is sound asleep within minutes, and she wakes refreshed around noon. It's dinner time back in Connecticut, so she calls Tim and the children before grabbing a meal in the cafeteria. Then she looks for Dave, who has been searching scientific articles for any pertinent information about the moons of Mars.

Dave and Heidi decide the news from Mars is important enough to send out as an International Astronomical Union circular. They choose their words carefully because they don't want to overstate the case, but they hope that other astronomers will be able to follow up while Mars is still near its closest approach.

They're both serious, but they smile a lot. The hard work that led to this circular is a perfect example of why they love science. The most surprising thing on this observing run was what they saw on those little moons, and they didn't set out to observe them at all. But because they had some time and the weather conditions allowed it, they looked at the moons of Mars and found something that no one expected.

Heidi and Dave carefully consider how they will communicate their results to the scientific community *(left)*. Heidi *(right)* consults her data notebook, part of a collection of notebooks she's kept since graduate school.

~ Changes Ahead

As Heidi leaves Hawaii, she is looking forward to observing Uranus with the BASS team another time. She's still excited about what changes she might see as Uranus approaches its equinox in 2007. She knows she will need a lot of careful measurements of a very dim planet, but the challenge makes the work all the more appealing to her.

An equinox is a time of transition, and in Heidi's life transitions have always meant opportunity. Now that she has reached what people usually call "mid-career," she is seeing signs of transition ahead in her scientific path. She is becoming much more than a gifted observational scientist and public speaker. She is now a leader in planetary science and astronomy.

In recent years, Heidi has been asked to serve on several important scientific committees. One of the most significant is planning the scientific objectives for the James Webb Space Telescope, which is scheduled to be launched sometime after 2011.

The James Webb Space Telescope is an orbiting infrared observatory that will reside nearly 1 million miles from Earth, where it will investigate the first stars and galaxies of the universe.

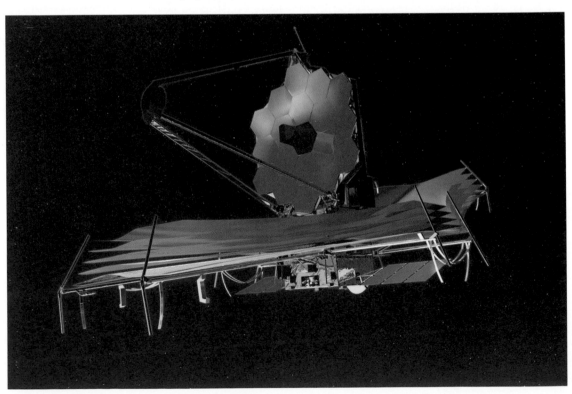

But now Heidi gets so many requests for her time that she needs to choose carefully among them. Most scientists would make their choice based on what the committee will do for their careers, but Heidi doesn't think that way. She decides by asking herself the same thing she has asked at every turning point in her life: Which choice is most challenging and interesting?

That question makes Heidi's future as unpredictable as the atmosphere of Uranus at equinox. Life's unpredictability has always meant opportunity for Heidi. The words of a Grateful Dead song describe it best:

Sure don't know what I'm goin' for
But I'm gonna go for it for sure.

No matter how much anyone plans, circumstances always bring unanticipated opportunities and challenges. The secret to success is recognizing opportunity, facing the challenges that it brings, and going for it—for sure.

Timeline of Heidi Hammel's Life

1960   Heidi Hammel is born on March 14 in Sacramento, California.

1970   The Hammels settle in Clarks Summit, Pennsylvania.

1978   Heidi graduates from Abington Heights High School in Clarks Summit.

1982   Heidi earns a bachelor's degree in earth and planetary science from the Massachusetts Institute of Technology (MIT) in Cambridge.

1982   While in graduate studies at the University of Hawaii, Heidi joins the choir of the Lutheran Church of Honolulu.

1983   Heidi's father, Robert Hammel, dies on July 18.

1984   Heidi travels to China as a percussionist of an English production of *The Phoenix Returns to Its Nest,* a Chinese opera.

1986   Heidi is invited to NASA's Jet Propulsion Laboratory (JPL) in Pasadena, California, to observe *Voyager* 2 imaging of Uranus.

1988   Heidi earns her Ph.D. in physics and astronomy from the University of Hawaii and accepts a post-doctoral position at JPL.

1989   Heidi is on the imaging team for the *Voyager* 2 flyby of Neptune in August. She also uses the University of Hawaii's 88-inch telescope to record red filter images of the planet during the flyby.

1990   The Hubble Space Telescope (HST) is launched into orbit aboard the space shuttle *Discovery*. Heidi finishes her post-doc at JPL and becomes a research scientist at MIT's Department of Earth, Atmospheric and Planetary Sciences.

1993   Eugene and Carolyn Shoemaker and David Levy discover their ninth comet in March. It is named Shoemaker-Levy 9. Heidi's proposal to study Neptune using the HST is accepted in December.

1994   In January, Heidi becomes principal investigator of the HST visible
       light imaging project for the upcoming Shoemaker-Levy 9 comet
       crash. In June, Heidi's work with the HST delivers an unexpected
       surprise—Neptune's Great Dark Spot has vanished. The Great Comet
       Crash takes place from July 16 to July 22.

1996   Heidi and Tim get married. Asteroid "1981 EC$_{20}$" is renamed "3530
       Hammel" in Heidi's honor. The American Astronomical Society
       awards Heidi the Harold C. Urey Prize in recognition of outstanding
       achievement by a young scientist.

1997   Heidi's daughter Beatrix is born on May 15.

1998   Heidi is presented with the Public Understanding of Science Award
       from the San Francisco Exploratorium.

1999   Heidi's son, Tobias, is born on May 27. Heidi later accepts a position
       as senior research scientist at the Space Science Institute (SSI).

2001   Heidi's son, Lucas, is born on May 22.

2002   *Discover* magazine names Heidi as one of the 50 most important
       women scientists. Heidi wins the Carl Sagan Medal for Excellence
       in Public Communication in Planetary Science.

2005   Heidi continues her work at SSI. She also speaks publicly about
       planetary astronomy and participates in educational outreach.
       She is a member of the team planning the scientific objectives
       for the James Webb Space Telescope, scheduled to launch in 2011.

ABOUT THE AUTHOR

After 25 years as a physicist and nine years of full-time writing, Fred Bortz is still trying to figure out what he wants to be when he grows up. That's why he loves to write about science and technology for young readers, sharing true stories about inventions, discoveries, and the people behind them. His work has earned him an American Institute of Physics Science Writing Award. *Beyond Jupiter* is his fifteenth book. Fred lives near Pittsburgh, Pennsylvania, with his wife Susan. They have two grown children and three grandchildren.

GLOSSARY

This book is about astronomy, the scientific study of stars, moons, planets, and other objects in the universe. The word *astronomy* comes from the Latin *astronomia*, which comes from the Greek word *astron*, meaning "star" and *nomus*, meaning "distribution." Many astronomy terms have Latin or Greek origins. For more information about each word, consult your dictionary.

asteroid: a rocky chunk of matter that orbits the Sun. Most asteroids, also known as minor planets or planetoids, are found between Mars and Jupiter. From the Greek *asteroeides*, meaning "starlike."

atmosphere: the layer of gases that surrounds a planet. Earth's atmosphere is made up of 78% nitrogen and 21% oxygen. The remaining 1% includes a mixture of water vapor, carbon dioxide, and other gases.

cloud: a collection of visible particles or gases. In Earth's atmosphere, clouds are composed mainly of tiny water droplets or ice crystals. The visible cloud tops of Neptune are thought to be made up of crystals of frozen methane gas.

comet: a small Sun-orbiting body, or nucleus, composed of icy rock, dust, and frozen gases. As a comet nears the Sun, heat vaporizes some of its frozen gases and releases some of its dust into a cloud around the comet. When solar wind strikes the cloud, it forms a long curving tail.

data: a collection of facts that can provide information or serve as the basis for arriving at a conclusion. Data can take the form of numbers or words or images.

eclipse: a phenomenon that occurs when one body in space blocks another, as viewed from a third body. On Earth, a *solar eclipse* occurs when the Moon blocks some or all of the Sun. A *lunar eclipse* occurs when Earth blocks sunlight and casts a shadow on the Moon.

equinox: an event occurring twice in a planet's orbit around the Sun, when all parts of the planet have equal periods of daylight and darkness. From the Latin *aequus*, meaning "equal" and *nox*, meaning "night."

galaxy: a large grouping of stars, dust, and gas that are bound together by gravity. Our Milky Way galaxy is only one of billions of galaxies that occupy the universe.

infrared: a band of the electromagnetic spectrum that lies between visible light at the red end of the spectrum and radio wavelength.

moon: the general term for a natural body that orbits another larger one; also referred to as a satellite.

observatory: in astronomy, a structure or site where scientists use telescopes to study objects in space. From the Latin *ob* meaning "in front of" and *servare*, meaning "to watch."

occultation: the blocking or dimming of a star's light that occurs when a nearby planet or asteroid passes between Earth and the star. This corresponds to an eclipse of the star by the nearby body.

opposition: a time when an object in space, such as a planet, is opposite the Sun relative to the observer.

orbit: the path that one celestial body takes around another, larger one. Such an *orbital* path is determined by the gravitational attraction between the bodies. From the Latin *orbita*, meaning "track of a wheel."

phenomena: facts, events, or circumstances that can be observed.

planet: a large, nearly spherical body in orbit around a star. From the Greek *planes*, meaning "wanderer."

probe: an unmanned space-exploration craft sent from Earth to gather and return data on objects or conditions in space.

rings: circular bands made up of rocky or icy fragments of matter that orbit a planet.

solar system: a star and all the bodies that are in orbit around it. Earth is part of the Sun's solar system, which also includes the other planets, their moons, asteroids, comets, smaller chunks of matter, as well as dust and gases.

star: a giant ball of hot gas that is held together by gravity and that creates and emits (releases) its own light from the energy of its core.

wavelength: the distance between one peak of an electromagnetic wave to the next.

Metric Conversion Chart

When you know:	Multiply by:	To convert to:
Inches	2.54	Centimeters
Feet	0.30	Meters
Miles	1.61	Kilometers
Centimeters	0.39	Inches
Meters	3.28	Feet
Kilometers	0.62	Miles

FURTHER RESOURCES

Women's Adventures in Science on the Web

Now that you've met Heidi Hammel and learned all about her work, are you wondering what it would be like to be a planetary astronomer? How about a wildlife biologist, a robot designer, or a forensic anthropologist? It's easy to find out. Just visit the *Women's Adventures in Science* Web site at www.iWASwondering.org. There you can live your own exciting science adventure. Play games, enjoy comics, and practice being a scientist. While you're having fun, you'll also get to meet amazing women scientists who are changing our world.

BOOKS

Kerrod, Robin. *Starwatch: A Month-by-Month Guide to the Night Sky.* Hauppauge, NY: Barron's, 2003. You don't have to board a spacecraft to explore the planets and stars. All you need is a view of the open sky and this book. Learn to identify the wonders of space, including comets, asteroids and other space phenomena. Sky maps and a pull-out "planisphere" will help you chart the stars from anywhere in the world.

Levy, David H. *Impact Jupiter: The Crash of Shoemaker-Levy 9.* Cambridge, MA: Basic Books, 2003. David Levy shares his amazing story of the discovery of the Shoemaker-Levy 9 comet (with Eugene and Carol Shoemaker) and its spectacular collision with Jupiter in 1994. This fast-paced, personal account is filled with all the thrilling details of one of the most spectacular events in our solar system.

Ride, Sally and Tam O'Shaughnessy. *Exploring Our Solar System.* New York: Crown Publishers, 2003. Take an exciting tour of the nine planets of our solar system with America's first woman astronaut. Get to know Earth and its neighbors, one planet at a time, through fantastic images and cool planetary and spaceflight information.

Voit, Mark. *Hubble Space Telescope: New Views of the Universe.* New York: H. N. Abrahms, 2000. No one can resist the astounding pictures taken by the Hubble Space Telescope. Some of the most incredible ones are here, along with the tale of this remarkable instrument and its contribution to our understanding of the universe.

WEB SITES

The Hubble Space Telescope: http://hubblesite.org/
The Space Telescope Science Institute in Baltimore, Maryland, hosts this official Web site, where you can witness the beauty of space through Hubble's eyes. Visit the gallery and then head over to the site's interactive "Fun and Games" page.

Infrared Astronomy: http://coolcosmos.ipac.caltech.edu/cosmic_classroom/ir_tutorial/
Learn more about infrared light (and other light waves in the electromagnetic spectrum) on this site run by the California Institute of Technology. Video animations, games, and image galleries reveal the mysteries of space—and of everyday sights right here on Earth.

Mauna Kea: http://www.ifa.hawaii.edu/mko/about_maunakea.htm
The Institute for Astronomy at the University of Hawaii manages the Mauna Kea Observatories. Here you can find out more about all of the telescopes that reside on the volcano's summit and even take an aerial tour of the area.

Space Science: http://spaceplace.jpl.nasa.gov/en/kids/index.shtml
Space Place is an excellent Web site from NASA that's jam-packed with lots of interesting word games, facts, and other hands-on activities covering a variety of space-science topics.

The Voyager Mission: http://voyager.jpl.nasa.gov
NASA's Jet Propulsion Laboratory is the place to go for all things Voyager, past and present, including mission news, facts, and images.

SELECTED BIBLIOGRAPHY

Burrows, William E. *This New Ocean: The Story of the First Space Age.* New York: Random House, 1998

Croswell, Ken. *Planet Quest: The Definitive Guide to the Scientific Exploration of the Universe and the Search for Life on Other Planets.* New York: Harcourt Brace, 1997.

Darling, David. *The Universal Book of Astronomy.* Hoboken, NJ: John Wiley & Sons, Inc., 2004.

Hoskin, Michael, ed. *The Cambridge Concise History of Astronomy.* New York: Cambridge University Press, 1999.

Levy, David H. *Impact Jupiter: The Crash of Shoemaker-Levy 9.* Cambridge, MA: Basic Books, 2003.

McCurdy, Howard. *Space and the American Imagination.* Washington, DC: Smithsonian Institution Press, 1997.

NASA TV. *Comet Impact 1994.* Set of twelve VHS tapes recorded on May 18, 1994 and July 16–23, 1994. Courtesy of Donald Savage, NASA.

Shoemaker, Eugene and Carolyn, interviews by Fred Bortz at the Shoemaker home and Lowell Observatory, Flagstaff, Arizona, on June 30, 1995.

Spencer, John R. and Jacqueline Mitton, eds. *The Great Comet Crash: The Collision of Comet Shoemaker-Levy 9 and Jupiter.* New York: Cambridge University Press, 1995.

INDEX

LIBRARY ADVISORY BOARD

A number of school and public librarians from across the United States kindly reviewed sample designs and text, answered queries about the format of the books, and offered expert advice throughout the book development process. The Joseph Henry Press thanks the following people for their help:

Barry M. Bishop
Director of Library Information Services
Spring Branch Independent School District
Houston, Texas

Danita Eastman
Children's Book Evaluator
County of Los Angeles Public Library
Downey, California

Martha Edmundson
Library Services Coordinator
Denton Public Library
Denton, Texas

Darcy Fair
Children's Services Manager
Yardley-Makefield Branch
Bucks County Free Library
Yardley, Pennsylvania

Kathleen Hanley
School Media Specialist
Commack Road Elementary
Islip, New York

Amy Louttit Johnson
Library Program Specialist
State Library and Archives of Florida
Tallahassee, Florida

Mary Stanton
Juvenile Specialist
Office of Material Selection
Houston Public Library
Houston, Texas

Brenda G. Toole
Supervisor, Instructional Media Services
Panama City, Florida

STUDENT ADVISORY BOARD

The Joseph Henry Press thanks students at the following schools and organizations for their help in critiquing and evaluating the concept for the book series. Their feedback about the design and storytelling was immensely influential in the development of this project.

The Agnes Irwin School, Rosemont, Pennsylvania
La Colina Junior High School, Santa Barbara, California
The Hockaday School, Dallas, Texas
Girl Scouts of Central Maryland, Junior Girl Scout Troop #545
Girl Scouts of Central Maryland, Junior Girl Scout Troop #212

Illustration Credits:

Cover Erin Kiernan Photography; **viii** Courtesy University of Hawai'i; **x** Richard J. Wainscoat; **1** NASA and Erich Karoschka, University of Arizona; **2** Courtesy Heidi Hammel; **3** Magrath Photography/Photo Researchers, Inc.; **4–5** © 1999, Calvin Hamilton; **6** Courtesy Heidi Hammel; **7** NASA/JPL; **9** Tom Jarrett; **10** (*l*) NASA; (*tl and r*) NASA and Erich Karoschka, University of Arizona; (*b*) Hubble Heritage Team (AURA/STScI/NASA); **11, 12, 13, 14** Courtesy Heidi Hammel; **15** Fred Bortz; **16** (*t*) Courtesy Heidi Hammel; (*b*) Francesca Moghari; **17** Courtesy Heidi Hammel; **19** Courtesy Michele Bendrick Jack; **20** (*both*) Abington Heights High School Yearbook, 1977; **21** Abington Heights High School Yearbook, 1978; **22** Courtesy Heidi Hammel; **23** Abington Heights High School Yearbook, 1978; **24** Courtesy Mary Rhodes; **26** Donna Coveney/MIT; **28, 29** Courtesy Heidi Hammel; **30** Courtesy Jim Elliot; **31** (*t*) Courtesy Adler Planetarium; (*b*) U.S. Geological Survey; **32** Courtesy Jim Elliot; **33** © Philip Gould/Corbis; **35** (*both*), **36** Courtesy Heidi Hammel; **37** Courtesy University of Hawai'i; **39** Rob Wood, Matthew Frey, Wood Ronsaville Harlin, Inc; Annapolis, Maryland; **40** (*t, both*) NASA/ARC; (*b*) NASA and Erich Karoschka, University of Arizona; **41** Courtesy University of Hawai'i; **42** Courtesy Kathy Crosier; **44** (*t*) Courtesy Heidi Hammel; (*b, both*) Courtesy Greg Poulos, www.psilo.com; **46** (*tm*) from *Chinese National Opera*, 1985; (*b*) Courtesy University of Hawai'i; **47** Courtesy University of Hawai'i; **48** NASA/JPL; **49** Courtesy Heidi Hammel; **50** (*t*) Courtesy Heidi Hammel; (*bl*) Heidi Hammel; (*br*) Richard J. Wainscoat; **51** H. Hammel (MIT), NASA; **52, 53, 54** NASA/JPL; **55** Courtesy Heidi Hammel; **56, 57, 58, 59, 60** (*both*) NASA/JPL; **62** NASA/JSC; **63** Courtesy Infrared Telescope Facility/NASA; **64** Courtesy Heidi Hammel; **65** NASA; **66** (*l*) NASA/MSFC; (*r*) NASA; **67** NASA/JPL, and MIT; **69** (*t*) © Peter Staettmayer; (*b*) Dr. H. A. Weaver & T.E. Smith, STScI/NASA; **72, 73** NASA/HST Comet Team; **75** NASA/STScI; **76** H. Hammel (MIT), NASA; **77** NASA/JPL; **79** AP/Wide World Photos; **80** Erin Kiernan Photography; **81** Fred Bortz; **82** (*both*) Courtesy Heidi Hammel; **84** Fred Bortz; **85** (*both*) Courtesy Heidi Hammel; **86** Darryl Watanabe, NASA Infrared Telescope Facility; **87** Andy Steere; **88** (*both*) Fred Bortz; **89** NASA; **90** Fred Bortz; **91** NASA; **92** Johannes Schedler; **93** (*both*) Fred Bortz; **94** NASA/STScI.

Illustrations: Max-Karl Winkler

The border image used throughout the book is a detail of a spiral galaxy, © 1996 Digital Stock Corp.

JHP Executive Editor: Stephen Mautner

Series Managing Editor: Terrell D. Smith

Designer: Francesca Moghari

Illustration research: Christine Hauser and Joan Mathys

Special contributors: Mike Brown, Meredith DeSousa, Brian Dewhurst, Mary Kalamaras, April Luehmann, Diane O'Connell, Mary Beth Oelkers-Keegan, John Quackenbush

Graphic design assistance: Michael Dudzik